Heredity Before Mendel

Heredity Before Mendel

Festetics and the Question of Sheep's Wool in Central Europe

First Edition

PÉTER POCZAI

CRC Press
Taylor & Francis Group
Boca Raton London New York

CRC Press is an imprint of the
Taylor & Francis Group, an **informa** business

First edition published 2022
by CRC Press
6000 Broken Sound Parkway NW, Suite 300, Boca Raton, FL 33487-2742

and by CRC Press
4 Park Square, Milton Park, Abingdon, Oxon, OX14 4RN

CRC Press is an imprint of Taylor & Francis Group, LLC
© 2022 Institute of Advanced Studies Kőszeg

Institute of Advanced Studies Kőszeg
Kőszeg, Chernel u. 14., 9730, Hungary

The present book is a revised, extended edition of '*A Festetics-rejtély: a genetika története és Festetics Imre hagyatéka*' by Péter Poczai published by the Institute of Advanced Studies Kőszeg in 2019 translated by Éva Szalai. Illustrated by Marcell Kismartoni.

ISBN: 9781032027432 (hbk)
ISBN: 9781032015088 (pbk)
ISBN: 9781003184973 (ebk)

DOI: 10.1201/9781003184973

Typeset in Palatino
by KnowledgeWorks Global Ltd.

Contents

Prologue

The words of my late mentor, evolutionary geneticist John Maynard Smith, still echo in my mind: *"It is very risky to trust morphological proof: it is much safer to rely on genetic evidence in biology."* With a slight exaggeration, we can say that genetics is the most serious discipline of biology. When I heard this sentence from JMS (sometime around 1990), I had not yet recognized that the foundations of the study of heredity (genetics) owe much to a Hungarian count who lived and created his life's work in the city of Kőszeg long ago: Imre Festetics. The fact that neither I nor JMS were aware of this clearly shows that all works which aim to patch this shameful gap in our knowledge have great significance. Readers can now access Poczai's book which, relying on the latest scientific findings, introduces them to the oeuvre of the originator of genetics as a concept and one of the founders of genetics as a science.

Why is genetics so important? Because it is the field where biology parts most conspicuously with the world of a purely physical approach. To ground this claim, a metaphor may help to strengthen the argument. In order to burn and propagate "spread" in some ways, a flame can be said to have a metabolism—many burned-out houses and wildfires bear witness to this fact. However, the traits of a flame are always determined by specific physicochemical conditions: by the time the flame spreads over to the third house, there is no trace of what caused the fire, whether a matchstick or a cigarette. Living organisms do not "spread" in this way. Living systems "spread" by carrying billion-year "memory traces" through their genetic subsystems. We should neither forget that Darwinian evolution would not be possible without heredity. Darwin was very lucky because he did not need to know the *mechanism* of inheritance in order to construct his theory. The only thing he had to consider was the *fact* of inheritance (like begets like), which Darwin had obviously recognized. He was aware of the achievements in artificial selection (breeding) as well as the common knowledge in the nineteenth century of how much people were ready to pay for the sperm of a breeding stud. They knew that *hereditary information is a valuable thing*. This was also clear to Festetics and his fellow sheep breeders, but it is also true, particularly for Festetics, that they might have learned something from Leibniz himself, following the dictum *Theoria cum praxi*. It is somehow amazing that the seminal work written by Festetics 200 years ago about the genetic laws of nature *(Die genetischen Gesetze der Natur)*; in his paper *On Inbreeding* was published in an economics journal *(Oekonomische Neuigkeiten und Verhandlungen)* in a town then called Brno.

It is worthwhile to have a good look at the painting depicting Imre Festetics. The sharp, inquisitive look, the slim figure, and the whole appearance radiating intellect remind me of a portrait of the young Newton. To lay the foundation of genetics, reveal the conditions for the utility of inbreeding, raise the possibility of what we now call mutation, and, finally, to point out that all this should be analyzed mathematically—well, it is more than enough for a contribution. It is also enough for us to *consider Imre Festetics the most significant Hungarian biologist of all times.*

Péter Poczai has paid tribute to the Festetics oeuvre repeatedly. In 2014, one of the most prestigious journals of general biology *(PLoS Biology)* published one of his articles, coauthored with research associates, which did so much for the belated recognition of Festetics. Let us briefly savor the title of this paper: *Imre Festetics and the Sheep Breeders' Society of Moravia: Mendel's Forgotten "Research Network"*! This very title shows *the modernity of Poczai's approach:* it is scientific and employs the perspectives of the history and sociology of science at the same time. The current culmination of this multidisciplinary approach is now in the hands of the Reader.

I would like to highlight the role of the Institute of Advanced Studies Kőszeg (iASK) since Poczai's book is a part of their program. He and I were both guest researchers at iASK, although in different periods. Péter Poczai's work is a very strong argument for the existence of such institutes. Today, when many people tend to measure the value of every activity by the profit it can momentarily yield, this kind of institute has an outstanding role in preserving culture because it allows us to go further and deeper in our inquiries.

A few weeks ago, one of the "last of the Mohicans" from the heroic times of molecular biology, Nobel laurate Sydney Brenner, died. Oddly enough, the Nobel Prize was awarded to him not for his groundbreaking work in deciphering the genetic code (matching the information in nucleic acids and proteins), but for the genetic analysis of the development of the small roundworm (*Caenorhabditis elegans*). I recall that, when he explained the molecular foundations of the genetic code to Kurt Gödel in Princeton (New Jersey, United States), the mathematician stood up and said, "That's the end of vitalism." There is a double twist here. One of them is that Gödel made this statement within the walls of the Institute for Advanced Study, the first of its kind to be founded (with prominent scientists, including Einstein) in the world. The other twist is that Gödel was born in Brno, where the Sheep Breeders' Society and Gregor Mendel, father of genetics, worked earlier. There, this is the "research network" Poczai and others describe in their article! They are right and fair to argue that Imre Festetics can be considered "the grandfather of genetics" from this perspective; thus we, biologists are his offspring in intellectual terms.

The genetic code is symbolic in its current operation, if not in its origin: it is a convention developed in the course of evolution that each DNA or RNA triplet is linked to a specific amino acid. Biotechnologists are gradually expanding this vocabulary in the hope of practical benefits. Natural language is symbolic too: the form of words is usually conventional (*fa, Baum, tree*). The symbolic nature of these two, utterly different, information systems is one of the conditions for the amazing creative power of genetic and cultural evolution.

Eörs Szathmáry

Author Biography

Péter Poczai is Curator and Associate Professor at the Finnish Museum of Natural History, University of Helsinki, Finland. He received his PhD from the Georgikon, University of Pannonia, Keszthely, Hungary. He is the author and coauthor of dozens of scientific peer reviewed journal articles.

Acknowledgments

I have to admit that this book has a long history that began in 2018 when I searched for unique island endemics and invasive plant species in the Greek archipelago. As a base, we found a house with a wonderful view on the island of Aegina, Greece, which proved to be excellent for retreating and concentrating on writing. In addition to the plants collected and manuscripts accomplished, an outline of this book was born, which only grew over the years before it finally took on its final form. Behind our house is almost in the center of the island of Aegina at the foot of the mountains, the ancient Eleonas, which we often visited. Unexpected, rare beauty, a place rich in flora and fauna. The undisturbed Eleonas, with its magnificent hundreds of years old olive trees, was a great inspiration for me conceiving this book. Greg and Eleni, thank you very much for your cordial hospitality. From the island of Aegina in the same year, I headed directly to Kőszeg, where I was lucky enough to participate in a seminar organized by the Institute of Advanced Studies Kőszeg (iASK). The symposium *"Genius loci"* that touched upon the life of Imre Festetics greatly inspired this book. I am also grateful for the generous support of this institute promoting the publication of this book. A year later in 2019, while I was a visiting research fellow in Radboud University, Nijmegen, I have completed the final draft of this book, which in its extended format is now available for the readers.

I also would like to thank several individuals and institutions for their help and support on many fronts: David Adamčík National Heritage Institute, Czech Republic; Michaela Růžičková and Jana Barinová from the Moravian Archive in Brno. I am grateful to István Bariska, Festetics's "voice" or, rather, "pen," for his beneficial work in transcribing the count's letters. Thanks also go to Jiří Sekerák (Mendelianum, Brno) for transcriptions and translations of (partly) burned handwritten documents prepared by Metternich's secret police. I also thank to Jorge Santiago-Blay for inspiring talks, and for arranging some of the translations prepared by volunteering translators of the Smithsonian Institution Hela Finberg, Rosanne Johnson and Anne Schwermer what

I was able to use for my work. Thanks to Lisa Muszynski for much help in revising the draft of this book. I thank my spouse, Aija Ronkainen, for her patience and understanding which allowed me to devote most of my time to writing this book. *Paljon kiitoksia! Köszönöm szépen!* At the same time, I owe my friends, colleagues, and family an apology for studying the history of genetics rather spending time with them. I thank you all for your understanding. Last but not least, I am also thankful to the British Society for the History of Science and the LUOMUS Trigger Grant supporting my work.

Introduction

Three hundred years ago the fundamental questions of "heredity" were quite obscure among biologists. Even today's discipline of biology emerged only in the 1800s. However, by the 1810s discoveries deriving from medical science and agriculture had revealed that there are some rules in the transmission of characteristics across generations.

One of the first steps in the series of these observations was that of a Hungarian count, Imre (Emmerich) Festetics (1764–1847) leading to the epochal studies of Mendel laying the foundations of present-day modern genetics and genomics. Recent research in the history of science sheds light on the fact that thinkers in the seventeenth and eighteenth centuries tried to understand reasons for the resemblance between parents and progeny, which they simply called "generation." Today the laws of inheritance belong to the general culture and it is hard to imagine why it was seen as a mystery in the past. However, only 300 years ago the word "heredity" had no biological meaning whatsoever. The explanation for this is really simple: there were no scientific findings to prove the existence of this phenomenon.

The basic phenomena of heredity are quite complicated to allow for exact interpretation through schemes that rely on *sensus communis* or the logic of Greek philosophy. Nevertheless, as we shall see, these "common sense"-based observations had brought forth the fundamental propositions of heredity as a concept. Like every basic truth, heredity also seems simple after it was understood. Such an essential relation became clear for us only as a result of the mutually supporting and enduring work done by numerous people. The basic questions of heredity had been recognized in the early nineteenth century, the scientific community had to reach an appropriate level to be able to analyze and interpret the data collected through experiments properly. On the other hand, society had to accept their findings, which required the separation of superfluous observations and religious prejudices (e.g., incest taboo) from scientific achievements (e.g., artificial selection). Therefore, the emergence of genetics as an independent scientific discipline in the twentieth century was a quite complex process and, as such, it can be interpreted only with an (interdisciplinary) approach crossing many fields and dealing, for example, with the flow of historical events, societal changes, and the spread of dominant philosophical and religious currents as well. This complexity and long prelude explain why the discipline had such a rough passage. To untangle the mystery of biological inheritance bold thinkers were needed to figure out the millennial puzzle of humanity.

Ongoing studies in science, history, and philosophy have revealed how Mendel arrived at his epochal and symbolical pea experiments. He was not a lone friar, but a trained natural scientist who could rely on the findings of hereditary studies often called as "hereditics" that had already been conducted for six decades in the Moravian capital of Brno and connecting Central European territories. In this city, four different approaches—biological, agricultural, practical, and medical—were synthesized in the first half of the nineteenth century. These disciplines provided a unique perspective for understanding the "innermost secrets of Nature." All this took place within a particular political context—quite peculiar for the Habsbug Empire of the time. As a result of this approach, "heredity"—a legal concept relating to property and monarchical power—gained a new meaning in both philosophical and practical terms.

This book aims to explore these linkages, although science, medical practice, and philosophy in the eighteenth and nineteenth centuries emphasized different values, which had a large part in isolating the idea of "the study of heredity" up until 1819, when these different strands could finally intertwine in Central Europe in Brno and Kőszeg.

1

The Enigma of Heredity

Ancient Theories of Generation: Seed and Matter

> Since Nature hath inviolably decreed
> What each can do, what each can never do; [...]
> For if the primal germs in any wise
> Were open to conquest and to change, 'twould be
> Uncertain also what could come to birth
> And what could not, and by what law to each
> Its scope prescribed, its boundary stone that clings
> *So deep in Time.*

Titus Lucretius Carus (98–55 BCE)
On the Nature of Things (De rerum natura)
1.585–587, 1.589–592
(Translated by W. E. Leonard)

The ideas we hold about heredity are as old as humankind. The earliest artifacts of human figurative art depict the female body or genitals, which are associated with fertility. Let us take the example of the Venus of Willendorf, with a height of only 11 cm, found by József Szombathy (1853–1943) in 1908 at Willendorf, Lower Austria, which gave the name of this archaeological find. It is now certain that people living in the Upper Paleolithic (30–10,000 years ago) did not aim to represent "Venus," the symbol of feminine beauty. To date, archaeologists can only guess the role of these artifacts representing nude women with crudely depicted body parts associated with sexuality, referred to as "Venus figurines." The statuettes always lack facial features and feet, which imply that they functioned as symbolic representations of fertility (Dixson and Dixson 2011). Female sexual features associated with

DOI: 10.1201/9781003184973-2

child-rearing are often exaggerated on these figurines, which also suggest that these small sculptures were linked to fertility.

Paleolithic people probably thought that women alone had the power of creating life through childbirth (Murray 1934). This assumption is also supported by descent-related terms used in communities that have survived from non-literate societies. For example, among the Papuo-Melanesians of New Guinea, the natives of the Trobriand Islands have no concept of biological fatherhood, which is reflected in their language. The word they use to refer to the father (*tama*) literally means the "husband of my mother" (Malinowski 1932). Practically, this term tends to denote the man in who's loving and protecting company the child has grown up. Other Polynesian communities associate female fertility with additional mysterious forces. The natives of the Marquesas Islands in French Polynesia attributed supernatural power to these forces, which can frighten away even gods and drive evil spirits out of the human body. During an act of exorcism among these islanders, a naked woman would sit on the possessed person's chest and used her pudenda to drive evil spirits out of the sufferer (Figure 1.1). African traditions also equate reproduction with the female body, believing that ancestralhood and heredity are based on the canonical understanding that newly born children are a continuation of diseased ancestors (Gable 1996). Every living being, person, animal, and herb, has an *orebuko* for the Bijago islanders of Guinea-Bissau (Gallois Duquette 1983). This ancestor's principle lives on after death, and only women have the ability to communicate with these spirits. As a result, motherhood is a core aspect of Bijago society, and every time a new life is formed in the womb, a woman comes into contact with the so-called "spirit world." Bijagos have developed a matrimonial culture as a result of their dominance over life (Henry 1993). But other cultures in Africa also associated fertility with the female body in a one-sided manner. During a ritual in ancient Egypt, women gave the gift of their genitals to freshly sown land in order to increase the harvest, which ensured both the abundance of crops and the absence of evil spirits.

We also know from ancient Egyptian mythology that, after failing to catch the goddess Isis, Seth ejaculated on arable lands, upon which his semen grew into wheat (Oden 1979). Similar stories can be found in the Bible. In the Book of Genesis, Onan was

FIGURE 1.1
Throughout the nineteenth century, there are numerous illustrations for Europe depicting a female figure as a symbol of fertility deterring even the incredibly powerful devil himself by displaying her genitalia. (Source: Illustration by Charles Eisen for *The Devil of Pope-Fig Island, Tales and Novels in Verse*. Vol. 2 London 1896, p. 130.)

not willing to enter into a levirate marriage with his widowed, childless sister-in-law, Tamar; instead, he "spilt his seed on the ground." This story is the source of the word "onanism," used to denote masturbation. It shows that the term "seed" was used to refer to both human sperm and the propagative part of plants.

The benevolent effect of the female genitalia on nature was also recorded by Caius Plinius Secundus[1] (23–79 CE) Pliny mentions the belief that if menstruating women go round the field naked, they can free the soil from all kinds of vermin and increase the expected yield.

Since the birth of human civilization, it is clear that contributions from both sexes are needed to create offspring. However, the degree of their contributions and that there is a consistent relationship between parents and offspring was not always so evident. It is a plausible assumption that the link between copulation and pregnancy was established with the spread of agriculture around 9,000 years ago. Mating in animals can be controlled by separating the sexes and castration. For domesticated animals, mating always coincides with ovulation, so there is a good chance that copulation will lead to fertilization. Hence, slowly, but with growing certainty, humanity recognized that the concepts of mating and generation are inseparable.

The prehistoric focus on women had been replaced by the male-centered view—which persists in much folklore today—that sperm or "seed," the only immediately apparent product of copulation, was responsible for fertilization. However, the appearance of particular quantitative traits, such as height, which often manifested in the offspring as a combination of the two parents, remained a mystery. For example, a tall man and a short woman had a child of average height. These observations gave rise to the theory of "blending inheritance." Ancient Greek physician, natural scientist, and anthropologist Hippocrates of Kos (460–377 BCE), who is also considered the founder of medicine,[2] believed that both parents produced "semen" or seminal fluids, which intermingled to create the embryo. Mixed traits were explained with this blending of male and female seminal fluids, and the sex and characters of children to be born were thought to be determined by whether the paternal or maternal "seed" became dominant during the mixing that followed copulation. It was thought to be the reason why some children inherited the mother's eye color, while others inherited the father's hair color. On the other hand, the intermingling of equal proportions of male and female seminal fluids[3] produces the traits of both parents in offspring. Hippocrates's theory of pangenesis was so convincing that it remained the dominant view among natural scientists until the nineteenth century. Thus, it also determined concepts of inheritance propounded by Charles Darwin (1809–1882).

By contrast, Aristotle (384–322 BCE) proposed a more tangible theory. He thought that a particle of our body cells was concentrated in the germ (gemmule), which would be ultimately incorporated into the gametes (Figure 1.2). Therefore, the germ contained all parts of the newborn body. Change in the reproductive matter actively contributed to the development of character in offspring. For Aristotle, the germs in female menstrual blood were formed into a new life by the movement of the paternal seed. According to his theory, the father's seed induced qualitative changes in the mother's reproductive matter, thus Aristotle was the first to recognize the fundamental female role in producing offspring. He also understood that particular unchanged characters are transmitted across generations. This theory constituted the basis of speculations in the ancient study of heredity.

Greek physician Galen of Pergamon (Claudius Galenus c. 130–200 or 210 CE) attempted to synthesize and rethink the theories of Hippocrates and Aristotle in a body of works that reflected the pragmatic Roman world view. Based on the theory of essential bodily fluids or humors, he divided people into four groups of temperament, which may seem to be ludicrous today. However, he recognized that certain phenomena of life and the laws that govern them can only be examined through experiments on living animals. Galen attributed great significance to the role of domestic plants and animals in studying these phenomena.

Aristotle and Hippocrates tried to unravel the process of generating life and explained similarities or differences between parents and offspring. Aristotle also understood that, as opposed to male-centered views of inheritance, maternal seed had a decisive effect on the offspring, just as plant seeds sown in different soils would produce plants with different forms. However, neither Aristotle nor Hippocrates articulated deeper relationships regarding heredity, since they could not find any consistency between the properties of parents and their successive offspring.

FIGURE 1.2
Aristotle, who in general had a greater role in the emergence of modern science, is often displayed in the seventeenth and nineteenth centuries as an elder man surrounded by various exotic species of animals and plants. The central figure of this illustration is a naked female body as a symbol of fertility or the source of Aristotle's seminal fluids. (Source: The Wellcome Library.)

Later, in the age defined by Christianity, St. Augustine of Hippo (354–430 CE) scrutinized the idea of heredity. His works tried to reconcile Aristotelian teachings with Christian theology and directives of the Old Testament which declared that the world came to be from nothing through divine creation (*creatio ex nihilo*). According to St. Augustine, God endowed matter with the disposition to self-organize, thus specific plants and animals can reproduce themselves to develop independently within the general order of nature (Figure 1.3).

While the ideas of Aristotle and Hippocrates dominated thinking about generation for almost 1,500 years in both the West and the Islamic world, differing views evolved on the other side of the planet, in China. In this part of the world, thinking focused on the "reproductive vitality" of men and women, defined in terms

FIGURE 1.3
The Christian theological belief of creation is displayed. The omnipotent holy tetragrammaton "Yahweh" is written in the center of the triangle, which symbolizes the Trinity. All species created by divine benevolence are depicted in pairs, with two unicorns in the center of the figure. While Adam and Eve are not yet present, a child can be seen in the forest. (Source: The Wellcome Library No. 15475i.)

of organ networks and their energy flows, rather than explaining the role of units in various phases of generation.

Generally, questions on how generation took place and what the contribution of each sex is in creating offspring were heavily debated in antiquity. It was even more obscure as to why progeny did or did not resemble their parents or why all this is possible at all. It is really disappointing to see that, ignoring the thoughts of ancient philosophers, humanity was not even groping in the dark in order to resolve the mystery of heredity: it had been relying on superstitions and religious dogmas rather than trying to understand this phenomenon for more than 1,500 years.

Modern Thoughts on Heredity: First Steps Toward a Scientific Approach

> For he who is acquainted with the paths of nature, will more readily observe her deviations; and, vice versa, he who has learned her deviations will be able more accurately to describe her paths.
>
> **Francis Bacon (1561–1626)**
> *Novum Organum*[4] (1620)

Synthesizing the advances in astronomy and physics, French philosopher and natural scientist René Descartes[5] (1596–1650) described the world around us in a new framework, through relationships based on the principles of mechanics. Scientists realized that nature can be explored and explained if we understand the system of constitutive particles. All living beings were seen through mechanical laws. Isaac Newton (1643–1727) proved the existence of gravitational force mathematically, which created a need for studying the structures of the universe, macrocosm, and, later, the Earth. It took only a short time for scientist to focus their attention on the world of the living, and they began to scrutinize what the mysterious "generation" (or procreation) really was.

Although the observations discussed herein came from the study of development, they are also closely related to biological inheritance, since it is important to understand that the word "heredity" had no biological meaning in the seventeenth and early eighteenth centuries. As Sandler and Sandler (1985) pointed out, at this time, the concepts of heredity and development were not distinguished clearly as describing distinct phenomena. For the scientists of this age, heredity constituted only a particular step within the infinite process of development; it did not occur to them that transmission processes can and should be studied separately. The problem was that, while these scholars did attempt to find links or resemblance between successive generations, whatever they observed seemed to lack consistency. For example, children born to the same family sometimes resembled one of the parents, at other times, both parents, or often none of them.

English physician William Harvey (1578–1657) proposed a theory that linked the origin of bird eggs to the mysterious concept of generation and heredity. He assumed that the mixture of unformed substance from both parents produced new progeny, and that fetuses were formed through gradual development and lacked the characters of adults at an early stage. Besides, Harvey was the first to describe the system of blood circulation accurately, and his interest in embryology was not limited to experiments in eggs: he continued research on deer embryos. He suggested that there must be some kind of "mammalian egg" too, and that all living creatures come from an egg (*ex ovo omnia*) (Figure 1.4). In search of this mysterious mammalian egg, he dissected dozens of does and hinds that were kept at the royal parks at Hampton Court. His findings were summarized in the book *Exercitationes de Generatione Animalium (On the Generation of Animals)*, published in 1651. In this volume, he described that deer embryos looked like little balls, which made him believe that he discovered some kind of "mammalian egg" during his dissection experiments.

This form of epigenesis contradicted the dominant view on the generation of living beings. In Europe, it was generally thought that barnacle geese (*Branta leucopsis*) grew on trees or that insects arose spontaneously from dirt. In fact, people also believed that, although rarely, women could give birth to animals or even monsters. For example, in the first half of the seventeenth century, the famous Flemish alchemist Jean-Baptiste van Helmont (1579–1644) published a recipe for generating mice via putting a dirty shirt, together with a handful of wheat, into a jar. Van Helmont claimed that after 21 days "a ferment being drawn from the shirt, and changed by the odor of the grain, the wheat itself

FIGURE 1.4
Rischard Gaywood engraved the frontispiece for Willam Harvey's *Exercitationes de generatione animalum* (1651), which depicted an egg in the hands of Zeus from which all types of living beings emerge. (Source: The Wellcome Collection No. L0006635.)

being encrusted in its own skin, transchangeth into mice." Strange ideas of the origin and evolution of life could not spring from the beliefs of the superstitious masses only at that time. In the 1660s, the Royal Society of London, then seen as the world's leading scientific body, discussed in minute detail how to produce vipers from wet soil. In this strange world, where mice could be generated from the mixture of a dirty shirt and a handful of wheat, it was impossible to argue about the related issues of reproduction and heredity. The very use of the word "reproduction" in relation to generation became widespread only in the second half of the eighteenth century; thus, even the simple study of this topic seemed to be a stillborn idea in a milieu where insects could just spring from the dirt. It was the unquestionable order of nature. Those who held different views were, at best, ridiculed, or, at worst, burned at the stake.

Significant advances in understanding generation came in the second half of the seventeenth century, through the work of

Regnier de Graaf (1641–1673), Francesco Redi (1626–1697), Nicolaus Steno (1638–1686), and Jan Swammerdam (1637–1680). We can also read about their achievements in the excellent book by Ropolyi and Szegedi (2000). Using theory, experimentation and dissection, these seventeenth-century researchers proved that female organisms produced eggs (*ova*) and that "like breeds like," including insects. A few years later, Dutch draper and amateur naturalist Antoni van Leeuwenhoek (1632–1723) rocked the world with his discovery of male gametes (*spermatozoa*) through a simple "microscope," although the term *spermatozoon* was coined subsequently by Karl Ernst von Baer (1792–1876), who in turn was the first to observe the mammalian egg. When Leeuwenhoek examined human and (mainly) rabbit sperm (Figure 1.5) through his microscope, he concluded that, with proper magnification, little spermatic animals or "animalcules" (*Samentierchen*) became visible. He thought that eggs serve only to feed and develop the resulting *preformed* fetuses. At about the same time, Harvey's contemporary Marcello Malpighi (1628–1694) studied the development of bird embryos and concluded that these preformed embryos contained organs from the very beginning. Epigenesis, represented by Harvey, on the one hand and Malpighi's theory of preformation on the other hand gave rise to the first debate in embryology. Initially, in the seventeenth and eighteenth centuries, preformism was the prevalent view, increasingly drawing criticism. Later, with improving observation techniques, the theory of epigenesis had been proved to be right. According to this theory, the initially undifferentiated "egg" does not contain the preformed organism, which develops only through successive steps of ontogeny.

Surprisingly, no one in the seventeenth century assumed that egg and sperm were complimentary elements within the same process and made equivalent contributions to the offspring. The scientists of this age had failed to reach this conclusion; instead, they had been debating, for almost 150 years, which component was the key factor, the egg (ovists) or the sperm (spermists). Spermists argued that, for producing new life, only the germ cell (sperm) was essential with the egg serving only as food, while ovists held just the opposite, arguing that the sperm cell simply awoke life already hidden in the egg. Later, in 1745, this controversy produced the concept of "reproduction" (Roger 1997). In order to understand the debate, which seemed to lead

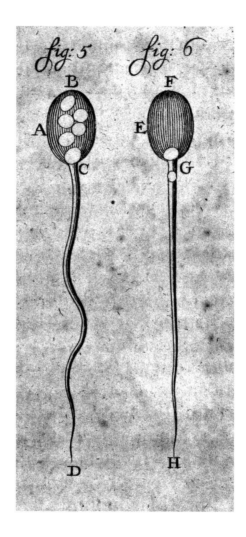

FIGURE 1.5
The illustration shows Leewenhoek's observations of rabbit sperm published in
Opera omnia, Vol. I (1722–1730), p. 168. The drawings display the body and structure of *animalculi*, with inner globulus particles and extending tails. (Source:
https://wellcomecollection.org/works/cabyhu72)

to a deadlock, we should examine the bases upon which the two
movements built their arguments.

A new approach to generation was introduced by French
zoologist René-Antoine Ferchault de Réaumur (1683–1757), who

studied the phenomenon of polydactyly. He assumed that reproductive matter contained the parents' organic molecules where,
after a union driven by special forces, a new order emerges, thus
the offspring is formed. The transmission of traits is linked to this
process. These ideas were further elaborated by mathematician
and astronomer Pierre-Louis Moreau de Maupertuis (1698–1759),
although the concepts of both French scholars remained little
known in the eighteenth century. In 1745, Maupertuis studied
albino children born with polydactyly. His observations led him
to conclude that both parents contributed "particles" to the offspring in equal proportions, and that their disproportionate
contributions would bring about "monsters." He thought that it
was the case for albinism as well as polydactyly and suggested
that an albino child's coloration could be due to a change in the
parental particles during transmission. This alteration was con-
spicuous, since all of the parents of albino children were black.

Maupertuis's theories were rejected by his contemporaries
because, adopting views that derived mainly from Greek atomist
notions, he dismissed both the ovist and spermist ideas mentioned
above. However, this could not discourage either Maupertuis
or Réaumur from proceeding with their studies of polydactyly
among the descendants of German and Maltese families (Glass
1947). Looking at the way this character reappeared over the generations, both of them arrived at conclusions sharply opposed to
preformationist theories. Maupertuis went even further when he
calculated the probability that this trait would not be transmitted
over the three consecutive generations, which produced the result
of 8×10^{12} to 1 (Sandler 1983). Up until the late nineteenth century,
the application of mathematics to the phenomena of biological
heredity could be considered a remarkable novelty.

Had Maupertuis proceeded to recognize further connections
and describe them with specific terms, the protagonist of this
book would have been this French scientist rather than Festetics
or Mendel.

Unfortunately, the discovery of parthenogenesis among aphids[6]
that Ch. Bonnet made with Réaumur's guidance in the 1740s
undermined his views on parents and traits transmitted through
their "particles," and he attempted no more inquiries into this
issue. How can anything be inherited if aphids can reproduce
without mating? Eighteenth-century scientists were deeply

divided into the ovist and spermist camps. The best example of this is that the Estonian Karl Ernst von Baer, who discovered the human egg, considered spermatozoa as parasitic worms. Natural scientists who studied generation seemed to reach an impasse in trying to interpret this phenomenon. Moreover, when the microscopic world opened up, they discovered still more unintelligible examples and inexplicable phenomena, which induced uncertainty and controversies. Therefore, the answer to the mystery of heredity was bound to emerge from agriculture with its much more practical knowledge, particularly from animal breeding.

The Roots of Heredity in the Intellectual Currents of the Enlightenment

Before proceeding to examine the effect of practical breeding on the concepts about heredity, let us evoke some of the intellectual and social transformations that occurred between 1685 and 1815 in Europe. The intellectual movement of the Enlightenment had a profound impact on and radically changed the then prevalent systems of politics, science, and philosophy—literature, poetry, and society alike. But it also had a great impact on members of scientific associations focusing on heredity through the ideals of Romanticism and German natural philosophy.

Enlightenment thinkers considered rationality as a governing principle and they questioned the arrangements of the world order, which affected institutional structures as well as established social customs and morals. They focused on rational thinking, as opposed to irrational thinking fuelled by superstitions (e.g., regarding the generation of mice employing a dirty shirt), which supported oppression-free and long-term development. Dozens of British, French, and other European thinkers turned against traditional authoritarian regimes, and the idea that humanity's progress depends on rational changes was increasingly espoused.

The Enlightenment produced a lot of books, inventions, scientific discoveries, and even laws, wars, and revolts. The American and French revolutions were also inspired by Enlightenment ideals, marking both its heyday and decline, which ultimately gave rise to the age of Romanticism in the nineteenth century. Important forerunners to the Enlightenment were Francis Bacon, Thomas Hobbes (1588–1679), and René Descartes, with Galileo Galilei (1564–1642), Johannes Kepler (1571–1630), and Gottfried Wilhelm Leibniz (1646–1716) being its key natural scientists. Most historians argue that this movement was rooted in the publication of Isaac Newton's *Principia Mathematica*[7] (1686) or John Locke's foundational text, *An Essay Concerning Human Understanding* (1689).

While Locke described the volatility of human nature and thought that our accumulated knowledge arises out of experience (empiricism), Newton provided a metaphor for achieving Enlightenment through his observations articulated in the principles of mathematics and physics. The political system of this age was characterized by enlightened absolutism in which, for example, king of Prussia Frederick the Great (Frederick II, 1712–1786) consolidated the state apparatus of Prussian military bureaucracy during the War of the Austrian Succession (1740–1748) and modernized Prussia. This was also the age when the Declaration of Independence (1776) of the United States of America was written. The Enlightenment had a profound impact on both religious and anti-religious movements. Christianity tended to define its religious doctrines on the basis of more rational principles, and believers got engaged in heated disputes with materialists over the origin of the universe.

Focusing on practical experience and advances in the natural science of their age, the thinkers raised their arguments against the dominant scholasticism and church authority. They tried to build a world on scientific foundations through seeking objective connections. Researchers increasingly immersed in the natural sciences, looked for causal relationships of a material nature, and formulated exact concepts and laws for distinct disciplines. Ordinary people turned their attention to nature and were followed by noblemen in their own way (Figure 1.6). An idyllic general image of nature emerged, and numerous aristocrats created parks with buildings constructed as picturesque venues for meeting with friends.

The spirit of this age is well reflected by the Hermitage near Dunkeld, Perth, and Kinross in Scotland, created by John Murray, Duke of Atholl, on the banks of the river Braan in the Craigvinean forest. Its building complexes include Ossian's Hall of Mirrors

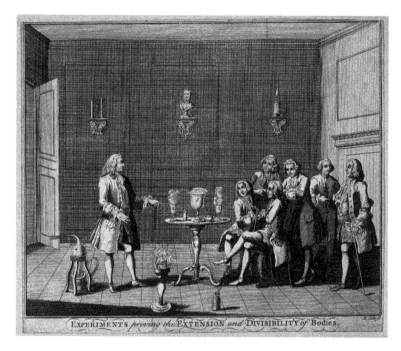

EXPERIMENTS *proving the* EXTENSION *and* DIVISIBILITY *of* Bodies.

FIGURE 1.6
A party of assembling nobleman listening to a seminar on natural and experimental philosophy proving the extension and divisibility of bodies. Engraving
by B. Cole (1748). (Source: The Wellcome Collection No. 46963i.)

and the hermit's cave, which gave its name. At that time nobles
had a habit of hiring hermits to occupy such hermitages who were
expected to live, for a specified period of time (usually a decade),
in caves, taking a vow of silence. They put these hired hermits on
show at social events and friendly gatherings, boasting of their
progressive thoughts about nature. Such jobs were not too popular though, since fees were payable only upon completion of the
service period. John Murray was left without a hermit because he
failed to find a "lucky" candidate for the position. Nevertheless,
these estates and country castles also hosted mushrooming
secret societies and gatherings, which were flourishing all over
Europe.

This age was also characterized in Central Europe by the
lack of universities and higher education in general, which self-
organizing, voluntary associations with scientific interest (Elliott

and Daniels 2006). These societies and academies constituted the backbone of the scientific profession. Their other important aim was to popularize science within a growing literate public.

Such associations sprang up in Moravia and Hungary too (E. Deák 2001), and were organized in Brno, later to become an important venue, by Count Karl Joseph Salm-Reifferscheidt-Raitz (1750–1838), the father of Hugo Salm, a would-be prominent player and key figure in this story. But members also included local nobles like the counts of the Mittrovský family as well as a Hungarian count, Imre Festetics in the period of 1787–1793 (Kühndel 1938; Kroupa 2006; Kodek 2011). Subsequently, these associations transformed into so-called "enlightened societies" for the cultivation of scientific culture. They organized themselves into *Vaterlandskunde* (fatherland study) movements, whose exemplars were the Agricultural Society of Brno (*Ackerbaugesellschaft*), to be discussed later, or the Sheep Breeders' Society (*Schafzüchtervereinigung*), with prominent members such as Hugo Salm, Christian Carl André, Johann Karl Nestler, and Imre Festetics (from 1814) or Cyrill Franz Napp (from 1826).[8]

In 1845, the young Gregor Johann Mendel joined this Agricultural Society, whose particular members with stronger interest in the natural sciences founded the separate "naturalist" section. From 1846, Imre Festetics and Mendel counted among its members at the same time (the former only for a short period, though, since he died in 1847).

Beginning in 1861, the section began its independent operation as the Natural History Society of Brno (*Naturforschender Verein in Brno*). The debates within the scientific workshops formed in the framework of this society reached two climaxes, which can be considered decisive in relation to the development of the study of heredity (evolving toward genetics). The first climax in 1819 featured Imre Festetics as the protagonist, while the second in 1865 had Mendel as its main character.

The intellectual bases of the Brno societies were determined by German natural philosophy (*Naturphilosophie*). This branch of philosophy attempted to conceive nature as an interrelated whole in order to define its basic unifying theoretical structure and lay the foundation of the natural sciences. Its representatives drew on the works of ancient Greek philosophers, as well as on the works of German idealist-philosopher Johann Gottlieb

Fichte (1762–1814) it was in the prime of the Romantic Age, during the first half of the nineteenth century, and strove to reach the unity of mind and nature. Its representatives saw the world and nature as a gigantic organism, which should be examined and scrutinized in order to decipher its innermost secrets (Figure 1.7). Johann Gottfried von Herder (1744–1803), for example, even considered languages as organisms which are continually formed, going through permanent transformation and changes. In 1772, Herder, in his holism-based linguistic conception published a *Treatise on the Origin of Language (Abhandlung über den Ursprung der Sprache)*. In his concept, the evolution of languages is a dynamic and never-ending process and all languages that can be traced back to this issue had common origin and have a "genetic" (*genetische*) relationship with each other. Another *Naturphilosoph*, the German poet and polymath, Johann Wolfgang von Goethe (1749–1832) suggested that plants can be derived from an archetype, due to a formative principle which originated all other plants (Brem 2015). He referred to the relationships of descent between plants as "genetic connections." These ideas became the basis for subsequent research conducted by Charles Darwin (1809–1882) and Alfred Russel Wallace (1823–1913), which later accrued scientific foundations to this approach (Wallace 1858; Darwin 1859). Both Herder and Goethe used the Greek word *genetikos* (γενετικός), in its original sense of "emergence, formation." Today's scientific terminology links the adjective "genetic" to the concept of the gene, for instance, in "genetic mutation" or "genetic disorder" (Szabó T. and Poczai 2019). For Herder and Goethe, this term denoted how different kinds were related to one another rather than biological heredity.[9]

The impacts of *Naturphilosophie* can also be discerned in the thoughts and views of the members of the Brno societies. It is no wonder, since the huge library of members included the foundational works of German philosophy of nature, and they presumably read those papers. They strove to study nature and to explore its hidden connections. They observed that "subtle problems are here to be solved before we can approach nearer to the truth … the innermost secrets of Nature" (André 1818). Thus, they adopted the motto of Festetics *"For your own part, work tirelessly, if you want to understand the rules imposed by nature upon itself"*.

Fitzgerald's General Preceptor. *Vol. pa.*

London Publish'd by C.Taylor Holborn Dec.r 12795.

NATURAL PHILOSOPHY.

FIGURE 1.7
Natural philosophy represented by a female figure experimenting with a vacuum pump. A bird can be seen dying the vessel. Stipple engraving by C. Taylor (1795), London. (Source: The Wellcome Collection No. 25995i.)

Notes

1. Pliny the Elder synthesized the scientific knowledge of his times in his thirty-seven-book *Historia Naturalis*. Following Aristotle's method, these books are based on observations and accounts confirmed by witnesses. The spirit of the age is truly reflected in the fact that, although he despised magic and strove to create a strictly scientific work, Pliny also described magical recipes and rituals.
2. Hippocrates and his followers were committed to the objective observation of the world. This school distinguished medicine, philosophy, and religion, which marks the beginning of a distinct medical science. Therefore, Hippocrates is also called the father of medicine, since he attributed all diseases to natural causes, omitting supernatural powers from life cycles.
3. The word *semen* is of Greek origin and it literally meant "seed"; in the English language, it is also used to refer to sperm, incorrectly.
4. Bacon's magnum opus, *Novum Organum* survived only in fragments. In this treatise the author was the first in the history of modern philosophy to propose the idea that inquiry should be guided by the (empirical) principle of experimentation rather than speculation. Therefore, Bacon clearly placed philosophy on natural scientific foundations, asserting that we can obtain reliable results only if we rely on facts, which must include the interpretation of nature.
5. René Descartes was one of the most renowned and most influential philosophers of the modern age. He determined the themes and approach of modern philosophy. As a practical natural scientist (a mathematician and physician), he was convinced that science cannot progress without proper support from natural philosophy. He strove to develop a metaphysics that could account for the possibility of scientific discovery.
6. Parthenogenesis is a form of natural reproduction which occurs vegetatively or asexually, and embryonic development takes place without fertilization. It is quite a frequent phenomenon in the living world, which can be found in invertebrates (e.g., roundworms, water fleas, scorpions, aphids, mites, parasitoid wasps, etc.) and vertebrates (e.g., fishes, amphibians, reptiles, and, rarely, birds) alike. It occurs much more often in plants and is called apomixis.
7. Its full original Latin title was *Philosophiae Naturalis Principia Mathematica*, translated into English as "Mathematical Principles of Natural Philosophy," which is often referred to briefly as the *Principia*.

8. *Vaterlandskunde* associations also focused on the natural sciences. They synthesized the characteristics of science, economic, and patriotic-humanistic societies, imbued with strengthening Moravian patriotism. At that time these associations had no political objectives. Most of them were philanthropic, humanitarian assemblages such as the *Humanität Gesellschaft*, which also took the role of learned societies in order to remedy the scarcity of universities.

9. It is worth noting here that this "Herderian sense" of the term "genetic" persist in colloquial language even today.

2

Sheep and Heredity

The Beginnings of the Agricultural Society in Brno

What makes a plenteous Harvest, when to turn
The fruitful Soil, and when to sow the Corn;
The Care of Sheep, of Oxen, and of Kine;
And how to raise on Elms the teeming Vine:
The Birth and Genius of the frugal Bee,
I sing, Mecaenas, and I sing to thee.

Publius Vergilius Maro (*70–19* BCE)
***Georgics,*[1] 1.1–6**
(Translated by John Dryden)

Interest in selective breeding and the subsequent socioeconomic changes during the eighteenth and nineteenth centuries demanded answers from the society of scientists. This offered an alluring opportunity for all those who were curious about resolving the secret of heredity, since the prompt practical exploitation of this phenomenon promised a fast track to wealth and social prestige. Therefore, the ambitious began to focus on numerous cultivated plants and livestock. Those who were committed to exploring heredity showed the greatest interest in sheep breeding because immense wealth and economic interest were concentrated in the wool industry. Research conducted in sheep breeding understandably provided the greatest thrust for inquiries into the enigma of heredity. Europe had several excellent breeders, with Robert Bakewell (1725–1795) of Dishley, Leicestershire, taking the lead by far. To say the least, Bakewell perfected sheep breeding and had the talent to increase animal growth rate and maximize useful tissue proportions based on minimum food intake

DOI: 10.1201/9781003184973-3

FIGURE 2.1
Robert Bakewell's barrel-shaped New Leicester (Dishley) sheep, created through inbreeding, as illustrated in Trousset's encyclopaedia (1886–1891).

(Figure 2.1). His achievements were acclaimed internationally and certainly induced a series of new debates and experiments regarding the question of heredity.

Unfortunately, history of science was never too enthusiastic about a theory derived from practice (*theoria cum praxi*) and usually looks upon breeders as people with wild ideas about inheritance. Personally, I share the opinion expressed by Dunn (1965): "*It was from experimental breeding that the principles of genetics were discovered.*" As we shall see, breeders tended to formulate their ideas in a simple and concise manner, an idea also shared by Darwin:

> What limit can be put to this power, acting during long ages and rigidly scrutinizing the whole constitution, structure, and habits of each creature,—favoring the good and rejecting the bad? I can see no limit to this power, in slowly and beautifully adapting each form to the most complex relations of life.
>
> **(Darwin 1859, p. 469)**

Darwin's clear, readable, simple but often eloquent style makes it evident that whenever breeders could promptly prove their theories with specific methods, they ensured their economic success and survival. It often required innovative methods and mindset, which defied the dominant social, religious, or even intellectual constraints.

Bakewell was quite secretive about his methods in print. However, he was ready to share these with disciples who arrived from almost all countries of Europe, including Hungary and Moravia. One of his most talented disciples was Baron Ferdinand Geisslern (1751–1824), also known in his homeland as the "Moravian Bakewell." Geisslern successfully employed the methods he learned at his estate in Hoštice, where he began to breed sheep. For this purpose, the Moravian city of Brno[2] and its surroundings, the hub of the wool and textile industries of the Habsburg Monarchy, proved to be a perfect setting. As opposed to Bakewell, Geisslern was not interested in increasing the quality and quantity of meat but in producing exquisite, fine wool. To achieve this goal, new theoretical and practical methods had to be developed.

To use a present-day analogy, Brno can be envisioned as an industrial city concentrating "state-of-the-art" methods and "cutting edge" technology that aimed to create an alloy of results from basic and applied research. To this end, efforts were made to accumulate significant know-how, represented by the methods of the famous breeding experts Geisslern and Petersburg. Their achievements were publicized by Christian Carl André (1763–1831), certainly the most committed scientist in Brno.

As a teacher, writer, and scientist, André had experience in different science disciplines and knowledge in economics. He aimed to promote the significant improvement of wool traits (e.g., crimp, density of staple, regain, etc.) and an increased yield of wool from sheep husbandry. He emphasized that this should be achieved through maintaining desirable traits in sheep stocks across generations. He was interested mainly in how climatic variations and nutrition affect wool quality, and how these variables influence the growth and reproductive capacity of sheep. Last, but not least, he also raised the question as to how these traits are transferred from parent to progeny.

During his 22 years of living in Brno, C. C. André was very productive in the publication and management of scientific efforts focused on these topics. The Sheep Breeders' Society

(SBS) was a truly "multicultural" scientific association, drawing a wealth of excellent thinkers and often audacious figures from various countries—including a prominent Hungarian count, Imre Festetics—and successfully gathered practical and theoretical professionals committed to sheep breeding. There can be no doubt that the "sheepy bunch" served as a kind of melting pot for combining different scientific and practical conceptions (Poczai et al. 2014). It should be noted that this intellectual context of Brno began to emerge earlier, with the self-organized Enlightenment societies (*Vaterlandskunde*) of the late eighteenth century.

How was it that this vigorous multinational intellectual milieu came to be developed in Moravia's capital city? Who were the prominent founders of the SBS? The *de facto* university of the Agricultural Society was active up till the end of the nineteenth century with around 300 founding members around 1820 that grew to approximately 8,000 by 1860. However, not all members were interested in heredity research, some of them were very enthusiastic about this subject. In the following I will list the biographical data for notable members who significantly contributed to forming lasting theories about heredity by being proponents of artificial selection, organizers, patrons, or they conducted experiments concerning the mysterious phenomenon of generation.

Christian Carl André: A Visionary Scientist of the Innermost Secrets of Nature

Christian Carl André (Figure 2.2) was born in 1763 and grew up in the Thuringian town of Hildburghausen, Germany. He studied at the University of Jena and made his name in Schnepfenthal first as a teacher committed to nature. Early on, André specialized in mineralogy and became a founding member of the Mineralogical Society of Jena. To supplement his income from teaching, he edited a series of encyclopedic volumes, reviewing and reading different fields of scientific endeavor with his associate and coauthor J. M. Bechstein (1751–1822). When editing the volume on zoology, published in 1795 (C. C. André 1795), André devoted special attention to "developmental history" (*Entwicklungsgeschichte*), a concept introduced into scientific discourse by German physiologist Caspar Friedrich Wolff (1734–1794).

FIGURE 2.2
Portrait of the agricultural expert and journalist Christian Carl André in 1819,
engraving by Blasius Höfel after Ant. Richter. (Source: Courtesy of the Moravian
Library in Brno No. Skř.2L-0091.228.)

As an early advocate and promoter of the idea of epigenesis,
Wolff believed that the developing embryo carried a hidden entity
transmitted from parent to offspring in the process of generation.

He also supposed that this transmitted thing expressed itself in a variable form, depending on the influence of climate and nutrition. New physical variants came into being as a result of environmental influences acting on the seed of the parents, both egg (ovum) and sperm (Wolff 1759). In order to prove his theory, Wolff took the case of Ethiopians and Europeans, asserting that "with climate it is heat and its digestive force that have a strong effect on the formation of variations" (Gaissinovitch 1990, p. 195) and "constancies appear in structures ... only because environmental conditions are common or remain the same" (Roe 1981, p. 131). He argued that the digestion of different nutritive substances was dependent on environmental temperature, which determined the skin color in successive generations—but his scrutiny was driven by scientific rather than racist motives.

Studying the doctrines of preformationists and epigenesis, André came down in favor of the latter and held that parts from disorganized matter gradually organized in the course of development. He believed that examples from both animal and plant crosses supported this argument and concluded that the transmission of traits from both males and female had to be equally important. André dedicated his next encyclopedic volume to the theory of "formative force" (*nisus formativus*) developed by a physician of Göttingen, Johann Friedrich Blumenbach (1752–1840) who believed that this innate force acted upon the developing organism to give form to its inherent properties, based on which development could proceed epigenetically (Blumenbach 1781).

André considered Blumenbach's term an organizing force that would be characteristically different among, as well as between different species, since it gave rise to their differing attributes during development. Due to the dissimilarity hidden in *nisus formativus*, a sheep and a dog differ in their physical appearance, although both of them are affected by the laws of epigenetics. This innate force also prevents, for instance, a chicken from turning into a sheep by following epigenetic laws only.

In 1798, André moved to Brno to take up a teaching appointment, expecting better opportunities in Moravia for developing his scientific interests, particularly in mineralogy. Presumably, André was invited to the city by Victor (also spelled Viktor in various references) Heinrich Riecke (1759–1830), a theologian from Stuttgart who studied philosophy at Tübingen University. As the first Protestant pastor of Brno, Riecke built a school and a church, and as a committed member of the late Enlightenment movement, he promoted the modernization of society. In the absence of universities, he aimed to foster education and the development of a learned society in Brno through private scientific societies. Riecke's efforts were in line with the central objectives defined by the Habsburg Monarchy in 1765 in order to promote the emergence of a learned society with high-level agricultural knowledge. With André's arrival to Brno, the pastor found the right man for realizing this firmly set goal because the then 35-year-old C. C. André had proper knowledge and ambition to perfectly fit Riecke's vision.

This teaching position met André's highest hopes. In the thriving industrial city, many naturalist-minded citizens organized themselves into self-educating circles and various natural science societies in order to enhance their knowledge and skills. For example, one of these groups was the Society of Friends of Natural Science and Knowledge of the Country (*Freunde der Natur- und Vaterlandskunde*), as an active member of which André could build a network of interested acquaintances and useful contacts. Within six years of his arrival in Brno, he succeeded in publishing a textbook on mineralogy (C. C. André 1804).

In this invigorating environment, besides his work as a teacher, André could also utilize his organizational skills, devoting his time increasingly to popularizing science or setting up new associations. Together with Riecke, he founded the Humanistic Society (*Humanität Gesellschaft*), functioning between 1802 and 1804 and having among its members Count Hugo Salm-Reifferscheidt-Raitz, in short, Hugo Salm (1776–1836), who also became a close friend to André through their shared interest in mineralogy. As J. Röckel wrote in his memoir about André, "[h]e is a passionate mineralogist who now devotes himself less to his teaching than, with painstaking endeavor, to his extensive scientific correspondence and scientific activity" (Röckel 1808).

Rudolph André: Promoter of Artificial Selection

Rudolph was born in Gotha, Germany, July 9, 1793, as the elder son of C. C. André. At a young age of 17, he began his career as a sheep breeder in Moravia as the pupil of Gesslern, which he later continued in Bohemia. In 1814, he took over the administration of the Raitz and Blansko estates, which belonged to Count Hugo Salm; a few years later he started to organize the larger estates of von Tischnowitz. His major interest as an active member of the Agricultural Society consisted of perfecting sheep-breeding by thorough attempts creating new races through inbreeding, outbreeding, and artificial selection. In his work, he attempted to perfect noble races of sheep through consanguineous mating, since he considered that these are the only means to propagate valuable traits in a pure state through generation from parents to progeny. Rudolph believed that unconditional inbreeding could lead to retardation or suppression (*Hemmung*) in or even to degeneration (R. André 1816, pp. 41–42). He also stressed that animals possess the natural capacity and the potential (Anlagen) to be improved for a higher perfection. He believed that such potential is natural and breeders only *"assist Nature to develop the extra potential towards perfection"* (R. André 1816, pp. 95–96). He was an active writer and published numerous books about agriculture and animal breeding (R. André 1815, 1818, 1820). He died from pneumonia at the early age of 32 on January 12, 1825.

Hugo Salm: An Adventurous Nobleman Promoting Scientific Development

C. C. André's life took a radical turn when he met Hugo Salm (Figure 2.3). Count Salm was an influential aristocrat, innovator, and entrepreneur who promoted the development of the textile industry in the city of Brno. In his nearby castle at Rájec (Raitz) he had a huge library of almost 59,000 volumes, mainly in French and German but also some in English. At the castle, he also built his own laboratory to conduct experiments on metal molding as well as dyeing and waterproofing industrial textiles. His library was rich in works on natural science and technical subjects, although his interests also extended to the arts and even to occultism and alchemy. Count Salm had been engaged in multifarious activities, for instance, he promoted the introduction of cowpox inoculations in Moravia in the early nineteenth century (Freudenberger 2003, p. 234). He based his methods on empiricism in the spirit of the Enlightenment, and prepared a vaccine from horsepox in the same period.

FIGURE 2.3
Portrait of the philanthropic count Hugo Salm benefactor and patron of the Moravian Agricultural Society in Brno. (Source: Courtesy of the National Heritage Institute, Czech Republic catalogue number: RA8349.)

In the eighteenth and nineteenth centuries, smallpox or variola was a very dangerous and contagious disease. All attempts to put an end to this pestilence had failed. The method of protection that originated in China and came to be known as "variolation" was perilous and risky. During this process people were artificially infected with pox through their skin (scarification), altering the inlet for the virus, thus ensuring that they would be immune to the disease for good. Let us imagine how dubious and dangerous this procedure could seem when humans did not have the faintest idea about immune response or the fact that smallpox is an ortho-poxvirus of 186 kbp, double-stranded DNA, and 302–350 nm size. Immunity was also hard to develop because variolated patients as well as people infected naturally could pass on the infection. The solution is attributed to English physician Edward Jenner (1749–1823), who discovered in 1796 that artificial infection or vac-cination with cowpox provides immunity against contracting an infection of smallpox (Esparza et al. 2017). The application of this dubious procedure in practice revealed considerable courage and foresight, and also demanded prescience from Salm to have faith in its success.

Besides vaccination, he was also interested in caving, and led exploratory excursions in the nearby region of the Moravian Karst (*Moravský Kras*) (Sedlářová 2016). In 1798, he left for Berlin to study the process of distilling sugar from beets (*Beta vulgaris* L., cultivar group *Altissima*), which he successfully introduced in Brno. During another excursion, he worked as a simple miner in Příbram (Bohemia) in order to learn the methods of extraction.

The most decisive event in Salm's economic aspirations and adventurous life was his visit to England in 1801. On this journey he was accompanied by one of André's close friends, the pharmacist Vinzenz Petke, who acted as a liaison in the acquisition and shipping of chemicals for the Brno textile industry. Their travel had strictly "scientific aims," to gather useful information for Brno's industrial output. Before setting out, Salm consulted a knowledgeable friend, Count Leopold Berchtold, who had already made several visits to England, thus he could provide useful guidance for the two travelers. Salm and Petke arrived in England hungry for information that they believed could benefit their own businesses and Brno's alike.

Their host was no less of a personality than the renowned botanist and natural scientist Sir Joseph Banks (1743–1820), President of the Royal Society. Salm had valuable knowledge and experience that he intended to share on the breeding of Merino sheep, which converged with Banks's efforts to encourage Merino breeding in England. Their encounter had brought forth mutual regard and a relationship of reciprocal advantage, although Salm harbored different plans. His determination led him to engage in a most profitable act of industrial espionage and theft. Banks did not realize that Salm, disguised as a worker, visited several English textile factories and secretly purchased sketches of machines required for wool production.

He smuggled the drawings back to Brno across the English Channel, providing an important impetus to the city's industrial development (Freudenberger 2003, p. 234). Brno soon became "the Austrian Manchester." To formalize his connection with Britain, in 1802 Salm hastily married a Scottish woman, Maria Josepha *née*

MacCaffrey Keanmore (1775–1836); in addition he went into business with an Irishman living in Brno. The count's wealth grew rapidly, thus he decided to hire an economic adviser. In 1811, this position had been fulfilled by none other than his friend, C. C. André. This provided André a solid income and allowed him to turn his attention entirely to the improvement of Brno's emerging textile industry.

Imre Festetics: Planting the Seeds of Hereditary Thinking

The Brno society and the city's vibrant intellectual scene had been frequented, almost annually, by a young noble from Hungary, Imre Festetics (Figure 2.4). Although the history of the Festetics

FIGURE 2.4
Count Emmerich Festetics, sheep breeder, grandfather of genetics, around the 1820s. (Source: Portrait painted by August Friedrich Oelenhainz. Town Museum, Kőszeg (No. 55.11).)

family is well known, Imre remained quite obscure.[3] Ironically, a portrait of this young noble had been discovered at an exhibition of historic clothing (Szabó T. and Pozsik 1990), while another one, possibly depicting the elderly Festetics, is still awaiting authentication (Szabó T. 2016). The mysterious count was born as the third son of Pál Festetics and Júlia Bossányi in 1764 at the family's castle in Ság (today Simaság). At that time, this had been one of the significant estates of the Festeticses. Of his siblings, György became renowned both as the founder of the first (continental) college of farming, the Georgikon in Keszthely (Hungary), and as a prominent figure in cultural history too, while his other brother, János, is less known historically. Imre's parents had huge estates, mostly acquired by his grandfather, Kristóf Festetics in the 1730s and 1740s. Kaposi (2016) discusses more in detail about the family's estate scheme and Imre's agricultural activities, emphasizing that the other, later called Dég branch of the family, was engaged in farming, mainly in the historic counties of Vas, Zala, and Somogy. The exact size of these estates at the time cannot be determined, but later land surveys imply that their total area could be approximately 300,000 Hungarian *hold*s (roughly 170,000 hectares). The Festeticses had major manors in Ság, Keszthely, Vasvár, and Csurgó. In the eighteenth century the family had a close relationship with the Habsburg Court through various state assignments and positions. The fact that, besides these major manors, they also had a castle in Vienna also reflects their affluence.

The traditional aristocratic upbringing took a somewhat unusual turn in Imre's life in 1782, when he became a soldier against his father's will. A military career usually provided a channel for advancement for young nobles, especially of the top echelon of society. Peace prevailed during 1782, thus military service did not harbor any overt risks. However, the end of the decade brought about the war against the Ottoman Turks, where Festetics suffered an injury near Bucharest, which put an end to his military career. In 1790, he retired from service and began to seek opportunities in the traditional way of life for the upper nobility. He moved to Kőszeg and, together with his brother György, set off on a journey to England, where he acquired state-of-the-art methods of agriculture and sheep breeding. Although the voyage had a great impact on the brothers Festetics, little is known of this important event.

FIGURE 2.5
Festetics Library, Hungary's sole aristocratic library preserved intact after the two devastating world wars in the Festetics Palace, Keszthely, with a collection of 80,000 volumes. Currently, it functions as a museum (Helikon Kastélymúzeum). (Source: Courtesy of 3dpano.)

Count Imre was not only fluent in at least four languages but also well-read and well-informed. He had a strong interest in the subtle relationships within nature, literature, poetry, and the development and interrelation of languages, with international correspondence on all these topics (E. Deák 2001). He had a library at his disposal in the family's Keszthely palace, then holding thousands of volumes, to quench his desire for knowledge. The Festetics Palace at Lake Balaton included a library so huge that a separate wing had to be built (Figure 2.5). This library has survived and is open to the general public. The collection included even works from Young, Culley, Sinclair, and Marshall (Kurucz 1990), as well as all significant journals and publications on the theoretical and practical foundations of agriculture and animal husbandry. Based on this accumulated knowledge, in 1803 Imre Festetics set out to breed sheep in Kőszeg and Kőszegpaty.

J. M. Ehrenfels: The Proponent of Climatic Upbringing

While we know little of Imre Festetics's life, it is a real riddle as to who Baron Ehrenfels was. Historians abbreviate his first name with the letters J. M., which stand for Josef Michael (in some sources) and Johann Markus, or the combination of the two, in others. Ehrenfels was an informed agriculturist of his age as well as a thinker interested in the arts and literature and a prolific writer on agricultural themes. He had successfully managed estates and promoted "the growth of fine wool" within the Monarchy as a whole. Relying on the work of Professor Thaer (1804), Ehrenfels (1817) emphasized the importance of selecting sheep and argued that the environment had a definitive impact on flocks. After the inbreeding debates in 1819–1821, he became formed his own ideas about the effects of the "genetic force" in the process of generation (Ehrenfels 1831). He was a propagator and defender of Professor Thaer's theses. The baron was also interested in animal diseases and apiculture. Together with Antal Nyáry (1803–1877) they summarized the rules and regulations of the Hungarian Sheep Breeding Society in Budapest (Nyáry and Ehrenfels 1830). From the 1800s until his death in 1843 he had been a member of the SBS of Brno and published several articles in its journals during this period.

Johann Karl Nestler: The Professor Investigating the Mysteries of Procreation

Nestler was born in the northern Moravian town of Vrbno, Czech Republic, in 1783. He studied law, theology, and jurisprudence at the University of Olomouc from 1800 to 1806, after attending secondary schools in Mikulov and Kroměříž. Nestler met C. C. André while completing his studies in 1806, and André offered him a position as teacher in the family of an Enlightened Baron D. Flick, at Staré Hobzí, South Bohemia (now Czechia). André also recommended Nestler for the administration of a private high school in Klasterbrunn, Austria, when he was entrusted with its organization in 1812. Nestler was in charge of the school until 1818, when it was ceased, then Nestler returned to the estates of Baron Flick where he was captivated by the progress of agriculture and animal breeding. Nestler arrived to Vienna in 1820 to study agriculture, his favorite subject, and where he was appointed as an Assistant Lecturer in the Department of Agriculture in September 1821. As advised to him by C. C. André after his departure in 1821 from Brno, Nestler became a Reader in Agricultural Sciences at the Francis University (Olomouc University) in 1823.

He then started lecturing about natural history the next year. Nestler received his Doctor of Philosophy degree from Olomouc University in 1828 and became the rector of the university in 1835 and Dean of the Philosophical Faculty in 1837 operating in Brno. Nestler maintained a membership in the Agricultural Society in Brno from the early 1820s till his death. It was this society where he met the proponent of inbreeding Martin Köller, with whom he began to collaborate closely, resulting in their shared approach to the advancement of knowledge behind sheep breeding success. They also benefited greatly from pioneering works of Geisslern and Festetics and both cited them in their papers and parsing the organizational efforts of C. C. André. Nestler's international recognition can be demonstrated by his nomination to membership of foreign agriculture and natural science societies in Wrocław (Poland), Karlsruhe, Darmstadt, Potsdam, München, Stuttgart (Germany), Gorizia (Italy), and Vienna (Austria). Nestler and C. C. André collaborated on a number of papers, and André was often called by Nestler as "an outstanding pioneer in scientific animal breeding." Nestler's contribution to animal breeding and his lectures about heredity, as well as his mathematical approach evaluating the passing on of sheep wool traits in six generations as envisioned by Festetics and C. C. André is often underappreciated. He's groundbreaking efforts are often unnoticed. Nestler's dedication to his work investigating the secrets of heredity left him with a number of health problems, and a pulmonary infection led to his premature death in 1841. Nestler was a well-known figure in Central Europe but also in the entire continent. Following his death, the secretary of the Agricultural Society, Lauer (1841), published an obituary note praising his contributions to Moravian research, especially to sheep breeding. About the same time but in Prague Hlubek (1841) also published an obituary but from a different angle highlighting his publications. An incomplete list was published in d'Elvert's (1870, p. 203) comprehensive monograph, where his papers are listed in chronological order.

František Diebl: Professor of Botany and Plant Breeding

Details of the early life of František (Franz) Diebl are unknown. He was born in Častolovice, Czech Republic, on August 8, 1770. According to d'Elvert (1870), he was born as a teacher and he had a strong collaboration with Nestler and Bartenstein arriving at the town of Brno for André's call after his departure to Stuttgart in 1821. He became a member of the Agricultural Society and for the suggestion of Cyrill Napp he was appointed as the secretary of the Pomological Society, and he also gained a position at the Philosophical Faculty of the Francis University (Olomouc University) together with J. K. Nestler (Weiling 1968). Diebl and Nestler tried to carry on with André's work and fill in the temporary vacuum generated by his departure by sharing dividing the task in a way that Nestler focused on animal breeding and Diebl concentrated on breeding of crop varieties. Diebl devoted himself to the establishment of a professional botanical collection by assembling a library with relevant literature and a botanical garden, which was opened to the public in 1828. Diebl was also devoted to popularizing new ideas and their implementation in agriculture for the propagation of fruit trees. In 1838, Diebl with Napp from the Pomological Society received a prize for breeding a new black currant variety, which found a practical response from M. Frey and J. Hard. Diebl was teaching supplementary agriculture of theological students at the Philosophical Faculty. The curriculum included growing fruit trees and viticulture. In the academic year, one of 1845/1846 one of his students called Gregor Mendel graduated from his classes and received a degree in practical agriculture. Diebl published his lectures from 1829 till 1844, where he concentrated on crop production specifically on artificial pollination of crop plants such as peas and beans (Diebl 1829, 1835–1841, 1839, 1844). Mendel learned from this coursebook and received the best possible mark on his examination. Diebl was a Moravian patriot and a proponent of the *Vaterlandskunde* ideology and struggled for an entire lifetime to make natural sciences a compulsory subject in general education across the Habsburg Monarchy. He passed away under an unknown condition in the town of Předklášteří on June 13, 1859. No portrait of Diebl has survived to identify him.

Cyrill Franz Napp: Patron and Benefactor Thinker

Franz Napp (Figure 2.6) was born on October 5, 1792, as the son of a shoemaker Ludwig Napp from Westphalia who came to Moravia as an Austrian soldier and later married Anna Maria Bergesin. Napp had a modest life and experienced poverty

FIGURE 2.6
Daguerreotype of Cyrill Franz Napp around 1841. (Source: Courtesy of the Moravian Gallery in Brno, Czechia, item No. MG15814.)

during his childhood, which eventually led him to join the Augustinian order in 1810 taking up his friar name Cyrill. He started working in Brno in 1815 as a scholar of the Old Testament and oriental linguistics. His career as a scientist was interrupted, when he was elected as the new abbot the St. Thomas Abbey in 1824. Unfortunately, because of his varying activities, he was completely withdrawn from his scientific work to which, he was devoting himself as a competent orientalist concerning the grammar of Aramaic and Arabic languages (Wurzbach 1869, pp. 81–83).

As an Abbot Napp, dedicated himself to the monastery's spiritual affairs as well as the management and improvement of numerous monastery properties. He eventually surrounded himself with a number of influential figures, including the polymath and family friend Johann Wolfgang von Goethe (1749–1832), the professor and later rector of the University of Cracow, F. Th. Bratanek. Napp held various honorable positions in public life in Moravia and Brno such as a member of the Moravian District Committee from 1859 to 1861. He was named as a permanent assessor in the Moravian-Silesian administrative district by the Austrian emperor in 1819. He was also instrumental in the establishing of the technical university in Brno, the creation of a professorship for Bohemian language and literature, the foundation of a savings bank for local businesses, and other important steps for the country's growth. He held numerous positions in various branches of the Agricultural Society as well. As an abbot he was responsible for the admittance of new members to the Augustinian order, Gregor Mendel benefited greatly from Napp's patronage for example appointment as a teacher, and not least in his university studies in Vienna. He also paid the inspection fees for Mendel at the Philosophical Faculty, and paid for the courses in plant breeding at the Agricultural Society.[4] Without any doubt, Napp contributed to the success of Mendel's work consisting not only in accepting the young Mendel in the religious community, who has no funding for further studies, but he also established a 250 m² garden space, and a greenhouse, which provided enough space for carrying out his experiments (Weiling 1968).

Gregor Johann Mendel: The Most Famous Member of the Society

Probably the most well-known member of the Moravian Agricultural Society (MAS) since he developed long-lasting theories about the field of science now called genetics (Figure 2.7). He was born in Hynčice near the Moravian-Silesian border on July 20, 1822, as the son of Anton Mendel and Rosine Schwirtlich (Klein and Klein 2013, pp. 91–103). He spent his childhood on the family farm and later worked as a gardener and beekeeper. He went to a gymnasium in Opava, and later continued his studies in philosophy and physics from 1840 at the Philosophical Faculty of the Francis University (Olomouc University). In Olomouc, Mendel had serious financial difficulties and he received support from his sister Theresa, but in September 1843 he joined the Augustinian friars and was assigned the name Gregor so he could finish his education.

However, we have no records detailing why Mendel joined the Augustinians; we can only speculate that it was because he wanted to find out how much he could expand his academic aspirations. As a result, he could dedicate himself to studies of the natural sciences for the benefit of his congregation. He was able to comprehend

FIGURE 2.7
Photograph of Gregor Johann Mendel. (Source: Courtesy of the Moravian Library in Brno No. Skř.1O-1156.035.)

that he had landed in a favorable scientific/artistic milieu, Mendel quickly understood the best approach for additional research. First, he devoted himself to nonacademic subjects, but spent most of his spare time studying natural sciences. After criticism Abbot Napp had to justify the scientific activities of the monastery in the presence of Bishop Schaffhausen in Brno. The authorities under the Holy Alliance intended to bring back the monastic institutions to Roman Catholic traditions. Abbot Napp prepared a document where he referred to the core pillars of the Augustinian order as intended to pass on Bible-based teachings and encouragement for the Philosophical Faculty. Along with botanists, mathematicians, meteorologists, and geologists, the other members of the monastery continued to operate in the same manner.

Most of the friars had taken membership in the MAS as well as further training at the Philosophical Faculty. Mendel similarly to other friars has taken up classes by Professor Nestler the head of the department and by Friedrich Franz who was responsible for the physics curriculum. Mendel's certification attests that he in fact earned top marks from the subjects taught by Diebl and Franz. With the approval of Abbot Napp, one of the fellow members of Mendel, Aurelius Thaler (1796–1843) who was a well-recognized mathematician and also a pioneer of botany, set up an experimental plot in the monastery's garden around 1830, where he grew rare endemic plant species. After Thaler's death, the small botanical garden was taken up by another friar Matthaeus Klácel (1808–1882) who was also a botanist interested in mineralogy and astronomy. With another friar Thomas Bratránek, Klácel often discussed about various philosophical subjects as well. Mendel joined the circle of these two fellow friar and Klácel as an older botanist became his mentor till Klácel was removed from his position as a teacher of philosophy despite the efforts of Abbot Napp, who tried to protect him Klácel was accused of pantheistic notions and he was termed as a Haegelian.[5] In an attempt, Klácel formulated a petition "in the name of humanity" that demanded more freedom for monastic orders and asked for the right to devote themselves entirely to science and education without prejudices.[6]

After Klácel's dismissal, Mendel finished his theological studies and went to work in a local hospital, where he became ill. Napp to show his support and sympathy sent him to Znojmo in 1849 where he was teaching mathematics in a high school. Mendel worked as a substitute before being sent to the University of Vienna to take the teachers' exam in order to become a full-time teacher. Taking such exams without prior study was only possible in Vienna at the time. Mendel took natural history exams, passing meteorological and geological subjects as well as mechanics, but failing zoology in 1850. After a year, Napp sent him back to the University of Vienna in 1851 to pursue a more comprehensive education in physics. In Vienna, his physics professor was Christian Doppler.

Mendel had no set courses in Vienna, but he was free to pursue whatever topic piqued his curiosity. Mathematics, chemistry, entomology, paleontology, botany, and plant physiology were among the subjects he chose. Physics, mathematics, and chemistry accounted for about 70% of his studies. He also took part in theoretical physics courses organized by Doppler. These courses had a major impact on Mendel, who was an excellent experimenter himself. Mendel assimilated much of the fundamental scientific knowledge that make up the modern scientist's worldview in Vienna. The physical aspect was the most prominent of this viewpoint, physics being the most evolved from all the natural sciences at the time. It taught that all aspects of nature obeyed certain laws, and that even the most complicated phenomena could be described by a limited number of laws dependent on the presence of the tiniest particles of matter. Science's aim was to learn more about these particles, discover the mathematical rules that regulated their actions, and develop a theory. Such rules should not be derived from speculative thinking or metaphysical phenomena, but rather from well-designed tests that could be mathematically validated and proven.

As I am trying to show in this book, all the elements required for preparing his study on heredity and developing his hypothesis were present in his university studies. In his hometown, the unanswered scientific topic had already been coined, and theoretical and experimental findings had been accumulating since 1819. In 1853, Mendel returned to his abbey to lecture, mostly physics. In 1856, he tried to become a licensed instructor but missed the oral section once more, and set to carry out his famous experiments with peas under the guidance of Abbot Napp. He succeeded Napp as abbot of the monastery in 1867. After being elevated to the abbot in 1868, Mendel's scientific work effectively came to an end as he became overloaded with administrative duties, especially a dispute with the civil government over its attempt to levy special taxes on religious institutions. He died of chronic nephritis on January 6, 1884.

Experimentation in Private Learned Societies

After his return from England, Count Hugo Salm contributed to the rise of wool production in Brno but, rather than keeping his profits from a growing industry, the committed philanthropist devoted his wealth to foster scientific research. In 1811, the enthusiastic count and his adviser C. C. André, possessing excellent organizational skills and an extensive network of contacts, cooperated in the fusion of the city's two distinct associations, the MAS and Natural Science Society to found the "Royal and Imperial Moravian and Silesian Society for the Furtherance of Agriculture, Natural Sciences and Knowledge of the Country" (Figure 2.8), hereafter briefly referred to as the Agricultural Society *(Ackerbaugesellschaft)*. In a speech delivered at the opening meeting of the new society, André envisioned a center that could serve both scientific and economic advancement, relying on the latest methods in experiments as well as practical experience, actively investigating and examining rather than simply observing, in order to give rise to the development of "the most useful auxiliary sciences" and encourage members to "pursue new ways,

FIGURE 2.8
Bishop's Court in Brno, housing the *Ackerbaugesellschaft* and the Moravian (formerly Francis) Museum. (Source: Image courtesy of the archive of the Mendelianum, Brno.)

better elucidated and richer in profit in agriculture and industry."
He argued that

> without science it is impossible to achieve any progress. [...]
> even though it may take centuries for works to emerge from
> our circle that are capable of earning the astonishment of
> the cultural world, and its gratitude for their public value.
> Whether today or tomorrow we are perhaps providing
> indispensable elements without even a hint of their future
> impact.
>
> **(C. C. André 1815)**

André transcribed and published his speech only years later.
The newly formed society had Hugo Salm as its president and
C. C. André as its secretary. Since sheep and wool provided a
basis for the welfare of Moravia and allowed Brno to flourish,
it was an important aim of the Agricultural Society to enhance
wool production through utilizing scientific achievements, thus
developing the economy of the city and the region. It is worth noting that the society, with some minor reorganizations, had been
maintained and working from its foundation in 1811 until as long
as 1898.

The members of the society recognized the economic benefits
offered by wool and accepted as fellow members renowned foreign experts of sheep breeding as well. Membership quickly grew
to 300–400, and the organizational structure of the society was
considered exemplary in global terms (Blum 1978, pp. 288–289).
Imre Festetics, who had an ongoing correspondence and active
contacts with André and Salm, also gained admission and began
to attend society meetings regularly.

It was by no mere chance that Imre Festetics began to attend
and became actively engaged in the meetings of the Agricultural
Society. In what follows, I briefly outline how scientific relations
between Hungary and Moravia developed during this period.
Eszter Deák (2001) gives a thorough analysis of this issue, arguing
that Czech/Moravian–Hungarian relations in cultural life and
the realm of science reached an unparalleled intensity in the era
of late Enlightenment, and we can find many similarities in the
historical and cultural development of the two entities/nations.
Such common foundations and parallels can be traced back to the

1770s and 1780s, the age of Josephism. This age can be characterized by the spread of interest in the natural sciences, with the Empire-wide close cooperation of scholars and active participation of learned nobles, in scientific life.

Besides concurrent cultural developments, interrelated efforts to achieve national independence can also be observed. The reform aspirations of diets held in the 1790s encouraged contemporary patriotic movements organized by the Czech estates, and the ideas of the Jacobin "conspiracy" in Hungary broadly resounded among the intellectuals of Bohemia and Moravia (E. Deák 2001). The library and museum founded by Ferenc Széchényi[7] and the Georgikon model farm established by György Festetics can be seen as further incentives that also served as examples for Czech patriots, drawing frequent references in contemporary press and correspondence. After 1800, especially from the 1820s onward, the national perspective strengthened on both sides; the pan-Slav idea emerging in Czech science on the one hand and the rise of the Hungarian national idea on the other hand necessarily tended to make an objective, nationalism-free, and mutually beneficial exchange of ideas between the two cultures, characteristic of the preceding period (a trend becoming later increasingly difficult to follow, cf. E. Deák 2001).

In lieu of a local university to centralize intellectual life, these societies were extremely popular and were quickly established in Brno—while the list of members often overlapped. The first pioneering association was established by Count Johann Nepomuk Mittrovský in 1770 even before the arrival of C. C. André to Brno and it was called the "Society of Agriculture and Liberal Arts" (*Gesellschaft des Ackerbaus und der freien Künste*). The second, established by Heindrich Friedrich Hopf and Count Hugo Salm-Reifferscheidt-Raitz, hereafter Hugo Salm (1776–1836) in 1790 was called the "Friends of Nature and Homeland Studies" (*Freunde der Natur- und Vaterlandskunde*). A third society, founded in 1794 and named the "Moravian Society of Nature and Fatherland" (*Mährisches Gesellschaft der Natur- und Vaterlandskunde*), was organized by Count Johann Baptist Mittrovský (1736–1811), Johann Nepomuk Mittrovský, Anton Friedrich Mittrovský (1770–1842), and Ignaz Mehoffer. Count Hugo Salm was not just active in establishing private learned societies, but he was also an influential aristocrat, inventor, and entrepreneur. He had a huge library of nearly 59,000 volumes in his castle in Rájec-Jestřebí, containing mainly French (e.g., Claude Adrien Helvétius, Jean-Jacques Rousseau) and English works (e.g., John Locke, Adam Smith).[8] His library was rich in science and technology, but his interests also included the arts and music, even occultism and alchemy. Hugo Salm's father Count Karl Josef von Salm-Reifferscheidt (1750–1838)[9] was a great admirer of the Enlightenment and had translated several works into German, including Montesquieu's *Considerations on the Causes of the Greatness of the Romans and their Decline*, or mathematical works written by Jean le Rond d'Alembert (1717–1783). His castle was the scene of both philosophical discourse and spiritualistic séances and alchemical experiments (McIntosh 1992, p. 67). In his spare time, Hugo Salm experimented in his private laboratory with metallurgy and investigated different dyeing techniques and aimed for waterproofing industrial fabrics.

As an active member of the Agricultural Society, André was able to establish useful relationships within these circles. In this intellectually stimulating environment, besides his teaching activities, he was able to leverage his organizational skills and devote more time to promote natural sciences and participate in the foundation of new industrial enterprises. Joseph Röckl, who visited André in Brno around 1800, wrote the following in his memoirs: [André] is a passionate scientist well oriented in different disciplines especially in mineralogy, and who is dedicating more time organizing science (Röckel 1808, p. 5). In 1800, André became a member of the Society of Friends of Nature and Homeland Studies organized by Hugo Salm. Soon André initiated the merger between their society and the one organized by the Mittrovskýs. This resulted in the creation of the Moravian Society of United Friends for the Advancement of Nature and Homeland Studies (*Gesellschaft der vereinigten Freunden zur Beförderung der Natur- und Vaterlandskunde in Mähren*) in 1801. During this period, André and Salm established a close friendship through their common interest and passion in mineralogy. They had several joint expeditions to the Moravian Karst (Sedlářová 2016, p. 6) and André published a full regional geological description where he attempted to explain the origin of the karst (André 1804). He also discovered a coal deposit near Rosice and, in 1803, he became the first in Brno to heat his apartment with it (Syniawa 2006, p. 27).[10]

Amidst the Habsburg Monarchy of the early nineteenth century, Moravians were not alone in their attempts at experimenting in diverse scientific fields or their efforts to connect people of the same interest. André was indeed one of the focal points in orchestrating such activities in Moravia, but many similar societies existed across the Monarchy. Indeed, the very organization of scientific activities generated heated debates across the Habsburg Monarchy from the mid-1780s to the mid-1850s. Several regions of the Empire presented their characteristic solution concerning the institutionalization of scientific activities. During the late eighteenth and early nineteenth centuries in the Habsburg Empire, research (*Forschung*) and teaching (*Lehre*) were rather distinct activities.

Research was pursued more successfully in private learned societies, where most of the members were amateur investigators and not university teachers. Experimentation was a popular activity, and, in consequence, many societies existed across the Empire. In founding such societies, they all issued practical and rigorous instructions for participants in how scientific work should be carried out; they characterized rules for observation, description, sorting, and classification (Kroupa 1998, p. 173). The constitutions of these societies strictly instructed the members to keep these rules within the context of their scientific work and to present their observations according to these practices.[11]

In Bohemia, Moravia, and Hungary, there were influential nobles who were doing research in their leisure time and achieved results similar to those of Hugo Salm.[12] Ignaz von Born (1742–1791) from a noble Transylvanian Saxon family was one of the notable experimenters with broad interests, including mining, mineralogy, paleontology, chemistry, and metallurgy.[13] In 1774, Born founded the pioneering Private Learned Society in Prague and he became the leading figure of the Enlightenment across the Monarchy. Members of such novel societies were eager to establish connections with well-known and prestigious scientific associations that provided feedback for the functioning of private learned societies and the work carried out by its members. Born was one of such members who gained a formidable reputation abroad for his work and, in 1771, he became a foreign member of the Royal Swedish Academy of Sciences; in 1774, he also became a Fellow of the Royal Society in London.

Based on his experiences in the Royal Society, Born suggested to the Emperor that the Monarchy should initiate scientific voyages like those of Captain James Cook (1728–1779). Born's success abroad inspired others to take up experimentation. For instance, Count Kaspar Maria von Sternberg (1761–1838) could be mentioned. Sternberg is often regarded as the "father of paleobotany" and established the Bohemian National Museum in Prague. He had a friendly relationship with both Johann Wolfgang von Goethe (1749–1832) and Alexander von Humboldt (1769–1859). The success of these members of various private learned societies across the Monarchy shows that, although the associations were formed by enthusiastic amateurs—who were able to shoulder the costs of their own research—later scientific experimentation found solid bases in their work. The newly formed societies attracted young teachers, research scholars, early career practitioners, and other interested individuals across the Monarchy—regardless of their social status—toward broadening their knowledge in various scientific subjects of natural sciences.

Breeding on a Scientific Basis

Both André and Salm were well informed about the new groundbreaking methods of sheep breeding, especially the ones that originated in England. In the first volume of *Patriotisches Tageblatt (PTB)*, one of the journals he edited, C. C. André published a favorable account of the techniques successfully applied by English breeders. Although he did not undersign the review, it was probably written by himself (Anon. [D. Le.] 1800). In 1804, the German translation of *Observations on Livestock*, a popular book written by Bakewell's closest disciple George Culley appeared that encouraged André to publish an article on animal breeding in *Hesperus* (Figure 2.9), another periodical under his editorial supervision (Anon. [probably C. C. André] 1809). *Hesperus* (named after the Evening Star, the planet Venus in the evening, "the shining one") reflected André's patriotic views, often allowing for criticism. Although it aimed to communicate the latest scientific knowledge and new intellectual trends for a knowledgeable public within the

Hesperus.

Encyclopädische Zeitschrift

für

gebildete Leser.

Herausgegeben
von
Christian Carl André.

XXX. Band. I. Heft,
(oder das 7te Heft des 13ten Abonnements). *)

Nr. 1 — 9. Beilage Nr. 1. Kupfertafel Nr. 1, und eine zu Nr. 6
gehörige Tabelle.

Prag 1821,
J. G. Calve'sche Buchhandlung.

*) Jedes Abonnement besteht aus 2 Bänden, jeder von 6 Heften, und kostet im Inlande 25 fl W, W.,
im Auslande 7 Thlr. sächs. Bei dem k. k. Oberpostamt in Prag pränumeriert man nur immer auf ein halb
des Abonnement oder Einen Band mit 16 fl. 30 kr. W. W. inclusive der postfreien Zusendung inner-
halb der k. k. Erblande.

FIGURE 2.9

Title page of Christian Carl André's journal *Hesperus, Education and Entertainment for Residents of the Austrian State* from 1821. Besides scientific announcements related to national enlightenment André also published entertaining fables and short stories based on legends and myths. One of his short stories was centered around Countess Elizabeth Báthory de Ecsed who has been labeled as a prolific mass murderer with sadistic tendencies. Her story describing Bathory's alleged vampiric tendencies bathing in blood has become part of national folklore grasped in André's (1817) paper. It also provided immense popularity for *Hesperus*, which become one of the most popular journals in the Monarchy based on the number of subscribers. Báthory's infamy still persists today providing cultural inspiration for the general audience.

Monarchy, it had certainly become the most prominent publication of Moravian liberalism.

Interestingly, André did not add his name as an author to this article either, only a footnote indicated that it was to appear in the 1805 volume of *PTB* but the publishing process ended abruptly due to the hostile reactions of the Habsburg Court to the ideas of Enlightenment. At that time Napoleonic Wars induced a hectic general mood. Since the publication of *PTB* had been suspended, André's paper could be published only later; while in 1811, he turned his attention to launching *Oekonomische Neuigkeiten und Verhandlungen (ONV)*, a periodical devoted to "economic novelties." Initially, *ONV* reported on the accomplishments of the Agricultural Society (*Ackerbaugesellschaft*) with a Europe-wide circulation of 6,000 copies printed in Prague and, like *Hesperus*, edited by C. C. André in Brno. Two of his sons, Rudolph and Emil became also involved in this work, the former preparing illustrations, including several blueprints for means of production and machines used in agriculture. It should be noted that the articles published in *ONV* were not confined to the theme of sheep breeding. As a journal specialized in agriculture, *ONV* discussed a wide range of topics from grapevine breeding or wheat production to animal and plant diseases. Besides acting as an illustrator, Rudolph also participated in research as an active sheep breeder and wrote a handbook on the subject (R. André 1816). Later C. C. André transferred his editorial responsibilities of *ONV* to his youngest son, Emil. The publication of these journals was financed by the bank of Brno.

ONV was advocated by several local patriotic nobles such as C. C. André's friend, Hugo Salm, or Leopold Berchtold, who had far-reaching merits in the advancement of Moravia's material and intellectual culture. In the beginning, the elder André's journal was committed to the Habsburg Monarchy, protecting the idea of Josephist Enlightenment, but some of its articles already reflected the ideals and values of early liberalism.

André's editorial approach was based on the ideal of English social and economic arrangements. It is no wonder, since Salm and members of the Brno societies also preferred to choose England as the destination of their travels with the purpose of gathering information or developing the economy of the region. Understandably, André considered this intellectual current and its ideas as exemplars. Among others, he drafted a plan for organizing Czech–Hungarian scientific cooperation, which is akin to the ideas proposed by Joseph Hormayr (1782–1848), another editor with links to Brno. The periodicals edited by André, *ONV*, *PTB*, and *Hesperus*, besides promoting the above conceptions, aimed to report on the events and novelties of economic and scientific life. These journals published many news concerning Hungary, which can be explained by André's interest in the country and its scientific world. Besides, these periodicals had a large readership in Hungary too. Their reports were regularly read by the counts of the Festetics family, and news on the Georgikon model farm and the experiments conducted there, with C. C. André's (even personal) supervision and evaluation in 1817, also appeared.

The articles published in these journals reflected André's concepts and knowledge of animal husbandry. This can be explained with the main objective of the editor, that is, to build the process of breeding on a scientific basis, thus pursue it at an advanced level. He viewed the practice of sheep breeding from a Silesian and Moravian perspective and highly esteemed the monumental work written by the famous Professor Albrecht Thaer (1752–1828), which was published between 1798 and 1804. The final of the three volumes on English agriculture was devoted to animal husbandry (Thaer 1804), and it was followed by a textbook on sheep breeding published several years later (Thaer 1811).

Thaer himself stated in this textbook that he made no claim to have extended scientific knowledge but wanted merely to summarize proven information already available on sheep breeding at the time. It was probably due to the pressure on him to get it written and published quickly. The members of the Agricultural Society had high expectations for Thaer's book and were disappointed by the lack of new ideas therein. This prompted Rudolph André to write a detailed and strongly critical review in *ONV* (R. André 1812). Besides Professor Thaer's book, André also launched sharp criticism, in his *ONV* reviews published in 1808 and 1813, on books about breeding Merinos by Swiss C. Pictét and French H. Tessier. Let us give an excerpt from the latter:

> If I were to rewrite my notes and arrange them in a unified account, a new book would result and, without exaggeration, one at least three times thicker than that of Thaer. The pick of foreign writings on sheep breeding, as for example

the ones I have before me by Mr Thaer, the one by Tessier, and the writings of Pictét,[14] fall far short of giving me a glorious impression of sheep breeding abroad. With what eagerness to instruct myself did I read these books! As great was my expectation, so was my dissatisfaction when I finally put them down. I particularly wished to find an answer to my doubts concerning the wisdom of crossing fine rams with common ewes—a controversial but common breeding practice. I have read Pictét's original French version, I thought to find everything on the subject, and I found almost nothing. [...] How could Thaer call his publication a "textbook"? A textbook ought to be comprehensive and exhaustive. The uninformed should be able to consult the books for advice. We possess in Moravia a school for sheep breeding such as the rest of the Empire and foreign countries can scarcely boast of.

(R. André 1813)

Following his father's advice, Rudolph André spent several months at Baron Geisslern's estate in Hoštice[15] to study the practice of sheep breeding. Inspired by his stay there, he began to draft a book on his experience of excellent methods and flocks. Meanwhile, in 1812, C. C. André serialized a paper in *ONV* by the Austrian Bernhard Petri (1767–1883),[16] who gave vivid and informative descriptions of his visits to sheep breeding farms in Spain, France, and England. In these accounts Petri drew parallels between animal and plant breeding, concluding that different races of sheep were equivalent to different varieties of plants (Irtep 1812).

Petri argued that both animal and plant breeders had a chance of being successful in selecting new varieties within the progeny of crosses. He was convinced that breeders would be able to "make these accidental varieties endure," provided that they could free themselves from any prejudice against inbreeding, It should come as no surprise that the application of inbreeding and crossing in novel ways originated in England, where experimentation with this innovative method was not hindered by religious sentiments. The matter of inbreeding[17] increasingly became a major discussion point at the meetings of the Agricultural Society. Sheep breeders began to scrutinize whether inbreeding had an unfavorable influence on continuity between generations.

Petri argued that the race of Spanish Merinos was maintained in its original form, as a "genetically fixed race" (*genetisch befestige Rasse*) due to climatic conditions. In line with dominant theories, he interpreted this in terms of "an internal prototype of organic formation" conditioned to act in such a way that if a reversion (*Rückschläge*) or a freak of nature (*Spielart der Natur*) appeared, it corrected itself in successive generations, regaining its main "racial form" (*Hauptgeschleschtsform*). However, he did not believe that the condition of the race was left entirely to nature. It is known from Petri's report of his visit to Spain that, although no spoken account or written source remained to support this assumption, the Spanish probably practiced selective inbreeding.

As a result, the enigma of generation, which determined scientific thinking on the study of development in the seventeenth and eighteenth centuries (see Chapter 1 of this book), became a central issue of breeding. Petri wrote an article on this enigma which was published by C. C. André in *ONV* (Petri 1813), in which he stated that "upon the generative substance hangs responsibility for fertilization and the progeny produced." I think that the gist of Petri's article is best summarized in the following lines:

> [...] when pairing two completely homogeneous animals of each sex, under favourable conditions, their inherent properties and superiority appear in the progeny, which are referred to as animals of the pure animal race.
>
> **(Petri 1813)[18]**

Therefore, the problem was how two completely homogeneous animals can be produced. Breeders in the Geisslern tradition thought that it was impossible, thus they preferred to maintain strict selection. C. C. André as an editor gave a platform in *ONV* to a variety of opinions about inbreeding and selection. In their contributions breeders increasingly focused on the nature of generation, speculation being organized around the following two questions:

1. What induced constancy in the internal organic structure of races, and what made them different?
2. What stabilizes this organic structure: favorable environmental conditions or, rather, selection?

The latter, the controlled selection, would imply that breeders could intervene in the form of newly generated animals through carefully planned selection. This was the principle that the two Andrés aimed to prove. They were convinced that the desired homogeneous herds of animals could be produced through artificial selection and inbreeding.

All in all, the breeders of Moravia were on the right path toward possessing the necessary theoretical knowledge and producing the best quality wool demanded by these times; in addition, this was the main reason why the idea of breeding on a scientific basis emerged within the circle of their Agricultural Society. Members began to propose thoughts that transcended practical breeding, which refutes the claim that Moravian breeders were interested solely in the special field sheep breeding. Perhaps they had a narrower focus initially, but then their debates on generation, the "formative force" shaping individuals and offspring laid the foundation for understanding deeper relationships. On Salm's suggestion, C. C. André published an invitation for "the friends and supporters of sheep breeding who, driven by patriotic sentiments, are interested in the advancement of the industry, particularly breeding," active within the Monarchy (Salm and André 1814). The aim of the two friends was to organize a fair where all breeders could exhibit their sheep, trade breeds, and to buy and sell new rams and ewes to improve their stocks.

The Taboo of Inbreeding

Central European breeders turned their attention to the methods of plant and animal breeding introduced in England, where, according to Joshua Lerner, the arrow of influence had led from agriculture to scientific progress (1992, p. 12). During the 1790s, bad weather and poor harvests called for an emergency to improve crop varieties in Britain. Thomas Andrew Knight (1759–1838) suggested that "climate" was largely unconnected to these failures and the cultivated fruit trees are coming to the end of their lifespan. Lidwell-Durnin (2019) argued that Knight's work played an important role in changing people's understanding about heredity disconnecting it from the rule of "climate" and concentrating on external factors. Breeders of the British Isles aligning with the machine age became aware of this phenomenon and produced popular new breeds of crops and farm animals. The New Leicester sheep highly bred by Robert Bakewell (1725–1795) is an illustrative example of how enlightened breeding can transform livestock into *"machines, for converting herbage, and other food for animals, into money"* (Sinclair 1832). To say the least, Bakewell perfected sheep breeding, and had the talent to increase animal growth rate and maximize useful tissue proportions based on minimum food intake. Bakewell's success was based on an excellent methodological approach: he pursued inbreeding (breeding in-and-in) in a closed stock (*abgeschlossene Rasse:* crosses between close relatives), which helped him decide that "seed" has a more important role than "climate" in shaping the animal's body form.

It seemed that, with consanguineous matching, there was a chance to fix by "blood" specific traits of animals. Bakewell earned great fame, both nationally and internationally, by effectively exploiting the power of "heredity" but his inbreeding methods were opposed on religious grounds. The well-known physician Caleb Hillier Parry (1755–1822) had the following statement about the degenerative effects of Bakewellian refinement:

> Breeding in-and-in [...] has been suggested to me by Mr. Davis, who thinks the early fattening of the New Leicester to be chiefly owning to this cause. He says that this constant incestuous intercourse produces, in both sexes, a deficiency of the powers of generation, and that of nursing in the female; reducing them to a state approaching to that of eunuchs ... If this opinion be well founded, it shows that the Divine Law against incest has a physical as well as moral end.

(Parry 1806, p. 471)

The quote from Parry shows very well that proponents of "Divine Law" rejected the consanguineous matching of sheep, which went against the cultural norm of the incest taboo; they were, however, bound to frameworks that were more traditional. Christine Lehleiter suggested that the acceptance of inbreeding required the separation of religious and biological questions, which was a step many were not yet willing to take (2014, p. 51).

Deviations from the "original form" of animals was understood as degeneration and such "freaks of nature" were not allowed to have a lasting influence on nature's overall design created by the omnipotence of God. Thus, inbreeding was related to the breakdown of sexual dimorphism causing infertility or even preventing copulation due to the lack of reproductive organs by reaching the "state of eunuchs." For those arguing against inbreeding, degeneration of sheep was a practical confirmation against human incest, which thus gained not only moral but also direct biological endorsement. Gottlieb Carl Svarez (1746–1798) the Prussian jurist and reformer drew examples from animal breeding in his lectures (1791–1792) to justify the moral objections against incest:

> The reasons for which marriages among close relatives are prohibited are partly physical, partly moral. The physical reasons are based on experience, which one has made with all kinds of animals that from the mixing of too closely related blood, races develop which—in particular when the mixing is continued through several generations—are marked very unfavourably by weakness stupidity, depravity.
>
> (Svarez 1960, p. 317)

Incestuous relationships were legally prohibited in German lands, and sheep conceived through consanguinity created in complete freedom from the bonds of matter were against moral standards (Figure 2.10). Christians adopted the sheep (lamb) as a religious symbol from ancient cultures and from Jewish sacrificial traditions, which reflected their understanding of how redemption is achieved (Harney 2004, p. 3). In this environment, the Moravian society was unwilling to accept the artificial modification of these symbols (Orel 1997, pp. 315–330). Around the early eighteenth century, only Ferdinand Geisslern (1751–1824), the "Moravian Bakewell," and Imre Festetics, the "Hungarian Geisslern" were applying inbreeding to improve wool quality in Habsburg territory.[19] While some experimenting and open-minded breeders in "harmonious unity" already put their knowledge into practice on the transformation of "noble races" from one to another, theoreticians and society were more sceptical about the "improvement upon Nature."

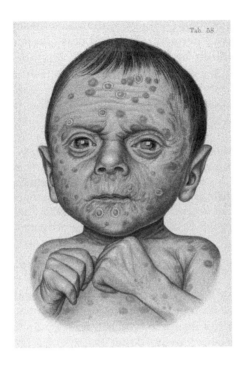

FIGURE 2.10
The bacteria *Treponema pallidum* causes congenital syphilis, which is transmitted from mother to infant during fetal development or at birth. The illustration depicts a baby suffering from such condition, where the head, shoulders, and hands are covered in pustules and keratitis. Since antiquity throughout the medieval times physicians played a crucial role in spreading and shaping the discussions about hereditary. Pathological appearances were perceived as monstrosities by the public while in a medical-discourse doctors observed that the occurrence of some diseases is not coincidental, they are often linked to certain families. They started referring to such disease as *morbi hereditarii* or hereditary diseases. At first there was no general distinction between truly hereditary diseases or infections passed on from mother to offspring at birth.

Consanguineous pairing was opposed by notable animal breeders and scientists debating over the adoption of the novel method. These included Franz Fuß (1745–1805), professor of agriculture at the University of Prague, who rejected inbreeding due to the potential harmful effects caused on the degeneration of the progeny (Fuß 1795); while Christian Baumann (1739–1803) a Cistercian monk, suggested changing the mating of rams every three years to avoid degeneration (Baumann 1785, 1803). The natural scientist

Johann Georg Stumpf (1750–1798) from Saxony supported these views and corroborated Georges-Louis Leclerc de Buffon (1707–1788), who in his *Histoire Naturelle* supported crossbreeding as a model for improving farm animals (Buffon 1760, p. 152; Stumpf 1785). According to Buffon, breeders could use the forces of nature in moulding animals to their purposes, but inbreeding should be avoided:

> In order to have beautiful horses, good dogs, etc., it is necessary to give foreign males to the native females, and reciprocally to the native males, foreign females; failing that, animals will degenerate [...] In mixing the races, and above all in renewing them constantly with foreign races, the form seems to perfect itself, and Nature seems to revive herself.

> **(Buffon 1753, pp. 215–217)**

The progressive Johann Petersburg (1757–1839) the manager of the sheep breeding farm of the Archbishop of Olomouc (Olmütz) following Geisslern's example was outraged by Buffon's views on inbreeding. He was convinced that Buffon's false ideas penetrated the mind of farmers so deeply that it will be hard to free them from the "enigma of inbreeding." Petersburg also circulated a pamphlet, later published by André, where he stressed his views against Buffon and asked professor Fuß to provide solid evidence for the harmful effects of inbreeding. In support of inbreeding Petersburg placed a versed inscription above the entrance of the Archbishop's farm:

> Where fresh herbs smell like balm
> Diligently airing the sheepfold
> Not confused by prejudice
> Mates for Mothers coming
> From Fathers, Sons and Brothers
> Bred healthy and noble lambs
> Which bloom through such wisdom
> This is what I have followed
> For several years.[20]

Köller later defended Petersburg stating that if the Archbishop himself allows consanguineous mating on his own farm, then questioning would this method be rejected by other breeders (K in Mähren 1811). André also joined the escalating debate by advocating inbreeding and sent his son Rudolph to study at Geisslern's farm in Hoštice (Hostitz); he later wrote a book about sheep breeding based on his experiences (R. André 1816). André's intention was to purge the minds of Moravian breeders from the Buffonian spirit (Figure 2.11). To achieve his goals, André began to explore "the enigma of inbreeding" with scientific thoroughness by publishing reviews on the method and its applications (Anon. [D. Le.] 1800).[21] These papers published from 1800 onward in his journal *Patriotisches Tageblatt* (PTB, Patriotic Daily News) stressed the advantages of inbreeding and tried to open the minds of the breeders and the public to accept consanguineous matching of sheep.[22] André was in a very difficult situation, since at that time the atmosphere was politically overheated by the Napoleonic Wars stemming from the unresolved disputes of the French Revolution.

In 1804, the German translation of George Culley's book "*Observations on livestock*", prompted André to publish an animal breeding article—mostly discussing the inbreeding methods of Bakewell—in 1809 with a footnote (Anon. [probably C. C. André] 1809).[23] According to this footnote, the original article was to be published in the volumes of PTB in 1805, but the publication process was interrupted by the sharp reactions to inbreeding. According to André:

> [...] the English have made the greatest progress, especially in animal husbandry, and through long-term observation and persistent attempts to discover the most correct, safest methods based on the laws of animal reproduction (*die Gesetze der thierischen Fortpflanzung*). [...] The application and usefulness of ennoblement (*Veredlung*) is based on the following experiences [...] that can be achieved in three ways: I) Inbreeding (in-and-in breeding) pairing individuals of the same race based on valuable characteristics; II) crossing or mixing different races; and III) avoiding further interference and ennobling only through inbreeding. This method is the fastest and most effective way of improvement. However, the prejudice against mating in close relatives must be banished with this refinement.
>
> **(Anon. [probably C. C. André] 1809)**

FIGURE 2.11
Sheep with congenital defects presented lithographs. The phenomenon hydrocephalus shown on top of the figure together with other defects were often observed by sheep farmers, who regarded such creatures as monsters. (Source: The Wellcome Collection No. 42216i and 42213i.)

The publication of PTB was suspended in 1805 and André published his paper in *Hesperus* [24] André reacted to the suspension of PTB by planning to leave the Habsburg Empire and to resign as secretary of MAS. He wrote the following to his friend Professor Julius Franz Borgias Schneller (1777–1833) about the incident:

> My Viennese observers, some of whom are very lazy make sure [...] I should be afraid that some things will never get into your hands, due to *endless espionage*. I was forced to leave since they withheld the food for my mind [...] and the Patriotic Daily News (*Patriotisches Tageblatt*) had to be stopped. I had no army to command and had no plans to violate the law so I decided to leave, but *the Emperor himself kept me back* and Count Lažanský, the regional chancellor, requested me in a very flattering official letter to resume my pen again. I made two conditions: first, for a more *liberal censorship* of my writings, which has rarely been fair to me; and secondly, for *the free admission of all books* sent to me from abroad as materials.

> **(Münch 1834, p. 335)**

This shows that debates about sheep breeding with scholarly deliberations on the question of heredity were deeply intertwined with philosophical and political questions in Brno during the early nineteenth century. The regional chancellor issued an official decree in 1806 on the easing of the imperial censorship against André's writings, which allowed him to continue his work in Brno. With PTB suspended André founded a new journal entitled *Economic News and Announcements* (*Oekonomische Neuigkeiten und Verhandlungen or ONV*), which appeared weekly and had more scientific and technical bases compared to PTB.

The Foundation of the Sheep Breeders' Society

The Hungarian count Imre Festetics, who followed his family traditions and had been engaged in sheep breeding since 1803, also read the invitation and proposal issued by Salm and André (1814). As he could not find better animals and rams than his own stock,

ILLUSTRATION 2.1
Imre Festetics travelling with his carriage from Kőszeg to the annual meeting of the Moravian Agricultural Society in Brno.

he decided to attend the new association's meeting to be held in May of 1814 (Illustration 2.1). Due to the surging demand for agricultural products and the resulting favorable transformation in the last decades of the eighteenth century, numerous landed nobles in Hungary, among them the Festetics brothers, Count György and Imre, set out to modernize farming at their estates (Lukács 2009).

On these foundations, in July 1797 György Festetics, accompanied by instructor Károly Bella and one of his students, established the first agricultural high school of the continental Europe, the Georgikon to promote the improvement of farming in Hungary. This institution was designed to serve as a higher-level school for farmers with the liberal approach of the late Enlightenment. Within this institute, soon six more (altogether eight) educational departments with different functions were formed, each of which with the responsibility, directly or indirectly, to train professional stewards (Hungarian *gazdatiszt*), originally for the family's own estates. The institution was inaugurated on August 23, 1801, by Archduke Joseph Anton of Austria, Palatine of Hungary (1776–1847) by a symbolic ploughing. In the same year, György Festetics had the small gymnasium of Keszthely upgraded to include five years, and established there a convictus[25] for young nobles, which was relocated to Sopron in 1808 (Csíki 1971).

György Festetics's life came to a turning point when, as an officer of the Graeven hussars, he signed a petition filed with the Hungarian estates' diet at Pressburg (Hungarian Pozsony, now Bratislava in Slovakia), which led to his falling out of favor with the Court (Cséby 2013). Following his retirement from the army, György Festetics returned to Keszthely in order to manage his debt-ridden estates. His last confrontation with the Court occurred in 1797, during the general assembly of Zala County, where he refused to vote for insurrection against the French. As a result, he lost his position as a clerk at the Hungarian Chamber, and the emperor also banned him from entering the Court (Cséby 2013). Festetics resigned openly from all of his political activities but insisted on implementing his goals regarding the education and cultural development of its estates. The Viennese secret police kept him under surveillance, hence he had to tread carefully but decisively in order to realize his visions. However, his caution drew criticism from several contemporaries, which, although it did not weaken his determination, became the source of numerous conflicts.

Editor C. C. André sympathized with György Festetics's views opposing the Habsburg Court. Since the Festetics brothers had lively relations with Moravian nobility and breeders, they were also active readers of André's periodicals, thus they were enthusiastic to respond to his call for a breeders' association. Imre Festetics gathered his best breeding animals from his Kőszegpaty estate near the Hungarian town of Kőszeg and set out on his coach to the meeting of the Agricultural Society Brno in order to forge stronger relations with local breeders and exhibit his own flock. His earlier communications certainly prove that he had been conducting novel inbreeding experiments for a long time. It should be the subject of further research how, by whom, and when, he was informed about these groundbreaking techniques. His motivation and methodology are still an enigma for science and agricultural history to date. Was it his own decision to apply these methods? Or was he encouraged to do so by *ONV* news? Or was his own in-depth knowledge of the literature on sheep breeding the key? What we can know for sure is that his travels in England had a great impact on his agricultural views.

Festetics started his breeding experiments around 1803 with a small flock of Merino sheep, and could boast of achievements that drew the attention of people gathered at the 1814 meeting of the Agricultural Society of Brno, where he could not buy breeding animals better than his own. The result was a real international interest and attention. All interested parties wanted to know Festetics's method, that is, how he succeeded in breeding such animals with outstanding characters. During the same Agricultural Society meeting, another noteworthy event occurred, namely the formation of the "Society of Friends, Experts and Supporters of Sheep Breeding." A long subtitle explained the goals of this society grounded "for the achievement of a more rapid and more thorough-going advancement of this branch of the economy and the manufacturing and commercial aspects of the wool industry that is based upon it," to which I shall hereafter refer in this book as the Sheep Breeders' Society (*Schafzüchtervereinigung*), or SBS.[26] In terms of organization, the new grouping worked as an independent section of the Agricultural Society (*Ackerbaugesellschaft*) as the first animal breeding association in the world. It was specialized only in sheep breeding, thus Imre Festetics found the best possible professional association to broaden his expertise. Its members were brought together by their shared interest in science practice, education, and the free expression of their views.

Based on his 12-year experience as a breeder, in 1815 Imre Festetics wrote his first article on the subject for the Hungarian weekly *Nemzeti Gazda* (National Farmer) edited by Ferenc Pethe (1762–1832). Pethe himself was a highly qualified agricultural expert with broad international experience, since he toured almost all countries of Europe between 1788 and 1796, after finishing his studies at the Reformed College of Debrecen. The article written in Hungarian by Imre Festetics was published under the title "A call on patriots eager to improve and amend sheep breeding" (*A juhtenyésztés' jobbítását és pallérozását óhajtó Hazafiakhoz*), appealing for the foundation of a Hungarian "Society for the Perfection of Sheep Breeding" (*Juhtenyésztést Tökéletesítő Társaság*), with Brno's SBS, established only ten months earlier, as a model. The result was the formation of the "Sheep Breeding Association of Vas County" (*Vasi Juhtenyésztő Társaság*, 1815–1820), the second professional organization for animal breeding in continental Europe. Festetics also mentions in an article that the initiator of this program was Dávid Chernel, whose activities he praised by referring to him as "our own Gaissler," as an equal to Moravia's Baron Geisslern (Festetics 1815a). In the same article, he reports on his own participation in the meetings held by the "Sheep-Wise Company gathering in the capital city Brno in mid-May" (*Május' közepén, Brno Fő-Városban összegyűlő Juhos Társaság*), adding that he was always accompanied by his "favorite woolly-fleeced animals" on these journeys, and bought breeding rams and ewes in Brno. In the same year he wrote a short report on the foundation of the Hungarian Sheep Breeding Association for *ONV* (Festetics 1815b). This was his first publication in this journal.

Soon after Festetics's report was published, young Rudolph André finished his book in response to the works of Thaer, Pictét, and Tessier, which disappointed him. The book was focused on "principles derived from nature and experimentation" (R. André 1816). Imre Festetics quickly reviewed and acclaimed this book. Meeting in Brno, Rudolph André became convinced that SBS members were able to produce Merinos of a quality that improved upon the Spanish stocks, citing Petri's account of experiences gained during his visits to Spain. Moreover, André's book was the first to detail the economic background of sheep breeding, "the business of improvement," used as a manual for the assessment and breeding of sheep at least until 1837. André also pointed out that, adapting to the environmental conditions of Moravia, the Merino sheep raised there were "transformed" and became increasingly uniform in the better and more processible quality of their fine wool.

André kept his promise regarding the book, being the first to discuss the need to apply scientific methods in breeding, which he could confirm with reference to SBS articles published in *ONV*. By "the application of science" he meant that predictable and reproducible breeding results can be obtained only on the basis of well-founded and clearly defined concepts, for example, the optical examination of wool quality. To further this aim, he developed a micrometer, whose blueprint was published in *ONV* (Figure 2.12). This device was designed to measure the fineness of wool fibers on a seven-grade scale. The younger André believed that the nature-given body form and build of sheep could be artificially transformed with various breeding methods. The term "law" (*Gesetz*) often turns up in the articles written by the two Andrés for *ONV* as well as in Rudolph's book. They argued that, for different breeders to understand each other, precise definitions and various laws should be articulated in order to make breeding more predictable.

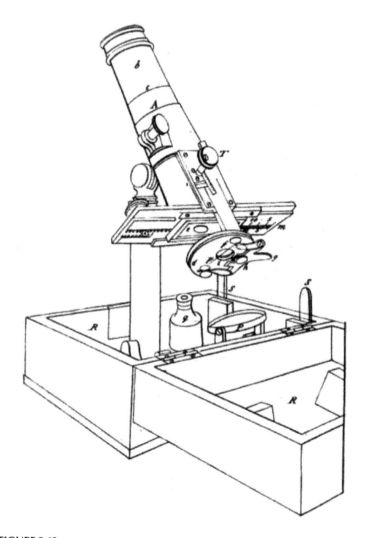

FIGURE 2.12
Rudolf André's micrometer presented in *Oekonomische Neuigkeiten und Verhandlungen* (Economic News and Announcements), 1821, No. 23. Vol. 29 Illustration No. 4.

Therefore, SBS members were encouraged to formulate their views on sheep ennoblement and inbreeding as clear and simple "laws," which could then be published in *ONV*. Baron Geisslern's associate Martin Köller (1779–1838) attempted to summarize "the laws and processes of nature (*das Gesetz und Gang der Natur*)" with moderate success, reaching the conclusion that "noble sheep without hereditary defects (*Erbfehler*), crossed with ewes without hereditary defects, produce offspring also without hereditary defects" (K in Mähren 1811). Only people with deep-rooted conservative prejudices, Köller believed, will oppose inbreeding. This refusal provided a suitable way for the lazy breeder to escape inconveniences. Köller never believed that inbreed groups of "noble" sheep would be perfect "finished products"; however, he proposed that they be subjected to continuous selection. *"The general rules of Nature cannot be modified only because it is easier to do so"* he cautioned. Inbreeding should only be practiced under rigorous selection, according to him, which would necessitate the breeder observing his stock animals with great wisdom and farsightedness, selecting desired characteristics from males and females and reproducing them in the future generations. These animals should be free of inherent flaws, which can only be accomplished by studying traits within inbred family lines of sheep, where certain defects can be discovered and sorted by the breeder. To accomplish this goal, several generations of the crossing would be required, and the breeder would need a thorough understanding of individual animals and their traits passed on from one generation to another in order to make the most successful crosses and remove unwanted imperfections. Any organism considered in this process had to be taken into account in its entirety, according to Köller, who also suggested the process of grading the animals generated by crosses, and he proposed what we would call today assortative directional selection with progeny testing as a method of doing so.

Through Practice to Theory

SBS debates held between 1816 and 1819 highlight that members participating in these meetings increasingly scrutinized how the traits of parents are transmitted to progeny. Understandably, they rarely used the word "heredity" (*Vererbung*), since neither the

ILLUSTRATION 2.2
Imre Festetics proposed the mathematical evaluation of wool fibres in sheep breeding. Rudolf André designed two micrometers and developed a system for the accurate grading of wool, which Festetics considered a milestone in breeding.

natural scientists nor the physicians of the age could explain the process of fertilization or the enigma of the embryo's origin and evolution. Hence, inheritance was a mystery that seemed to be inseparable from the process of procreation or generation (*Zeugung*), which produced offspring in the course of embryological development (Illustration 2.2).

SBS discussions combined the practical approach with theoretical questions. At the 1817 meeting, Austrian breeder Baron J. M. Ehrenfels warned SBS members that a regrettable reduction in the quality of the wool-fiber fineness was being experienced throughout the Monarchy (Appendix 1). Based on his best theoretical and practical knowledge, he perceived the following reasons for the degradation of wool quality:

1. the practice of selecting sheep for breeding mainly according to their physical appearance;
2. a neglect of wool quality in favor of quantity;
3. matching sheep "in closest consanguineous relationship."

According to Ehrenfels, animals produced by "incestuous" crossing (inbreeding) deteriorated the health of the stock as a whole (Ehrenfels 1817). He claimed that close inbreeding led to "natural climatic degeneration" by disrupting "the principal plasma of the animal's organization" (*Hauptplasma der thierischen Organisation*). In this early stage of the debate, Ehrenfels already linked the inconstancy of the animal's physiology with environmental impacts. To determine this issue, he asked for the opinions of other SBS members, admitting that "[n]ot every master of mathematics has to be a Newton, not every sheep breeder a Buffon, nor every collector of herbs a Linnaeus" (Ehrenfels 1817).

In response to the textile industry's request for clarification, the elder André published 36 questions to review the agenda of improving wool quality. Of these, he highlighted the influence of environmental variation on wool quality and the transmission of individual traits from father or mother to offspring affected by the "genetic force." The idea of "heredity" (without uttering this word or using this concept explicitly) was implied by several members, which had also been recorded by an anonymous reporter (Anon. 1818).

The 1818 SBS meeting was rich in developments, since Baron Bartenstein and Count Festetics, in opposition to Ehrenfels's opinion, defended the inbreeding method. In line with Rudolph André's views, they thought that pure Merino lines could be produced through inbreeding, and the purity of these lines could be maintained or even improved. As an example, Festetics introduced his Mimush sheep, what he has inbred for over15 years. In response, Ehrenfels took a stand at the other extreme, claiming that inbreeding had quite the opposite effect, and actually caused the degeneration of the stock, thus Mimush must show the signs of climatic setback and debilitation. To prove this, he relied on the example of Spanish Merinos. According to his views, the constant quality of their fine wool was determined by the Spanish climate, experienced over many generations. As opposed to Festetics's opinion, Ehrenfels claimed that outside Spain these sheep were liable to degeneration, reflected in declining wool quality. Thus, two different views emerged in relation to how wool fineness should be improved. One stance emphasized the importance of environment and feeding, the other gave primacy to internal factors.

The elder André was fully aware of the fact that it had become an established practice among breeders to inbreed mother to son and father to daughter. However, he also knew that they primarily carried out crossings with rams from different stocks and also relied on breeding rams from different flocks. André believed that, first, they had to agree on what they meant by "inbreeding" and clarify other imprecise concepts in their terminology too. In his related article (C. C. André 1818), he was in agreement with Ehrenfels in concluding that unconditional close mating of sheep of the same blood carried on for several generations must result in organic weakness, which he called "the physiological law of nature" (*physiologisches Naturgesetz*). In his opinion, inbreeding should be practiced with great caution before this method could bring promising results. Prior to further debates, in-depth studies should be conducted, whose results could be summarized in a book, because, as he concludes,

> subtle problems are here to be solved before we can approach nearer to the truth... [for] we are penetrating into the innermost secrets of nature.
>
> **(C. C. André 1818)**

The Genetic Laws of Nature

In order to clarify the disputed taboo of incestuous mating, André called upon Festetics to summarize his views in response to Ehrenfels. Festetics stated that such difficult questions require extremely precise definitions and deeper knowledge, which cannot be made without careful preparation (Anon. 1818a, p. 298). Therefore, he suggested preparing his responses in writing for the next meeting. The result was a longer paper, which was published in three volumes in the pages of ONV edited by André and appended by his extensive footnotes and a separate editorial note (André 1819; Festetics 1819a, b, c). Festetics defined himself as a "curious empiric" who has "gathered practical experience," what he has complemented with "occasional reading in natural history," where he did not find any answer to the "theoretical part" of Ehrenfels's argument (Figure 2.4). He admitted that the points listed by Ehrenfels could be true from a "purely physiological (*rein physiologisch*) point of view." But he was asked to "illuminate the meaning of the system that he has formed" to mitigate any concerns about consanguinity (Festetics 1819a, p. 9). Festetics formulated his explanations under five points of fundamental laws about organic functions (*Grundgesetze der organischen Funktionen*)":

1. I associate organic weakness [...] with the following definition: the subject in an otherwise good state of health is unable to perform and maintain its organizational functions in accordance with natural laws (*vermöge Naturgesetz*) for a relatively long period of time.

2. I include among the organic functions everything that the laws of nature obviously require from the subject to preserve its self-organization (*Erhaltung seiner selbst*) and to propagate it in subjects resembling to itself.

3. Robust constitution is related to the preservation of self-organization, which is partly inborn (*theils angeboren*) and which may partly increase or decrease by upbringing (*durch Erziehung*).

4. Precisely this robust constitution is necessary for the emergence of healthy entities (*Wesen entstehen*) resembling their ancestors in the process of reproduction (*Fortpflanzung*). Healthy fathers often produce less appropriate offspring (*erzeugen*). Thus, the constitution, regardless of the state of health may weaken.

5. If traits (*Eigenschaften*) that I desired for my purposes are fixed in the constitution of the Mother and the Father

and variation appear in the offspring, these are either
freaks of nature (*Spiel der Natur*) or the ancestors were
not adequately equipped (*hinlänglich ausgerüstet*) with the
required traits.

(Festetics 1819a, pp. 9–10)

He tried to answer whether any subject arising from consanguinity agrees with natural law or "lies outside of nature's bounds."
Does consanguinity prevent physical entities from integrating
their organic functions? By this, he meant the conservation of
self and reproduction of offspring resembling their ancestors.
He specified that the growth and development of entities depend
on environmental responses, which together with inborn components alter the structure and composition of entities. Stable
inner conditions, as Festetics explained, "robust constitution"
is required for the entities to reproduce healthy progeny, which
can deteriorate regardless of their state of health. But what if both
parents exhibited healthy constitution and had been carefully
selected to possess the desired traits? His answer was that even
in these cases variation (*Änderung*) could appear in the progeny
that he called "freaks of nature" or "sports." In his last sentence,
he also maintained the possibility that parents may not possess
the desired traits sufficiently enough to transmit them to their offspring. In a footnote, André added that there must be an error in
the transcript of this part of the sentence. Could it be that Festetics
meant here that "inborn components" of the parents must match
with each other in a specific way so that the desired trait physically
appears in the progeny? Festetics admitted that these explanations
are not exhaustive because "here we are only striving to search for
the truth" and the contradicting issues are only "verified by pure
experience" (Festetics 1819b, p. 19) (Appendix 2) (Illustration 2.3).

In the final chapter of his paper entitled "About inbreeding"
(*Ueber Inzucht*), Imre Festetics investigated whether this method
has a harmful effect on generation by breaking down the transmission of traits through degeneration, or that it leads to the
contrary state, more certain heredity? In a practical consideration,
does inbreeding debilitate sheep to a certain state when they
cannot mate and lose their "noble" characteristics or the exact
opposite, that they yield better, more refined wool? According to
the "humble opinion" of Festetics, the following four paragraphs

ILLUSTRATION 2.3
The inbreeding debate of the Moravian Agricultural Society.

contain the "Genetic Laws of Nature (*die genetischen Gesetze der Natur*)":

a. Animals of healthy and robust constitution plant and bequeath their characteristic traits

b. Traits of the predecessors, which are different from those of their descendants appear again in future generations.

c. The animals which have possessed the same suitable traits (*angeeignete Eigenschaften*) through many generations can have divergent characters (*abweichende Charaktere*). These are variants, freaks of nature, unsuitable for propagation when the aim is the heredity of desired traits (*Vererbung der Eigenschaften*).

d. Scrupulous selection of stock animals (*Stammthiere*) is the most important precondition for the successful application of inbreeding. Only those animals possessing the desired traits in abundant amount can be of great value for inbreeding.

(Festetics 1819c, p. 169)

André in a footnote added to the term "scrupulous selection" specified that "In my opinion, this underlines the main point."

In the first law, Festetics linked heredity with health and a robust constitution. Existence was tough for a breed imported into a strange environment as pointed out by Ehrenfels. Degeneration was an ever-present danger. "Noble" blood would not be transmitted readily by a sick male, nor could desirable lambs be expected from a weak female. In the second of his laws, he assured his fellow members that when a character skipped a generation this need not be considered a sign of degeneration. Such gaps in heredity were commonly observed and offered no barrier to eventual breeding success. The changes he referred to in the third law were of a different nature, deviations from normality that had to be excluded from the bloodstock. Such abnormalities, freaks of nature, might arise from a variety of causes, possibly connected with deviations from health and fitness referred to in the first law and noted in the fifth point of his "Laws of Organic Functions." The fourth and most significant law referred to mating among chosen bloodstock from which abnormalities had been expelled. To inbreed in such circumstances, each trait being separately considered was the way to maintain high quality. It was also, in some cases, a way to further improve the stock. The traits to be considered were not only those relating to wool quality but to health, fitness, and fertility (Szabó T. 2009).

Festetics believed his "organic and genetic laws" to be evident of daily phenomena in nature underpinning fundamental functional processes of "natural history," which proves that the method of inbreeding cannot be defined as "manipulation against ennoblement" that "goes against natural law" stated by Ehrenfels. How could Nature act against herself? Festetics verified his laws based on his observations and experiences in sheep, horse, goat, swine, horned cattle (*Magyar Szürke*), and poultry breeding (Festetics 1819a, p. 10; 1822, p. 729). Aligning with the spirit of the Industrial Revolution, he noted that the "manifold architecture of the horse machine" is harder to comprehend, thus inbreeding of horses is more challenging than that of other animals. Rudolph André in a separate paper verified Festetics's observation and added that only "homogenous animal races (*homogenen Racethieren*) possess the necessary organic strength" to produce the "noble race" and "consanguineous mating is the only available means by which the propagation of valuable traits in a pure state can be achieved in the progeny" (R. André 1819,

p. 161). He also repeated his previous expression that "such animals possess the natural capacity and the potential (*Anlagen*) to reach higher perfection assisting Nature" (R. André 1816, pp. 94–96).[27] In a separate editorial, C. C. André asked, "how many homogeneities and heterogeneities does the nature of wool have?" and "is it just a fantasy to ask if the same analogy of natural law could occur in the plant kingdom?" (André 1819, p. 29) (Appendix 3). Festetics's answer was that "among plants, fertilization of the female flower is achieved with the flutter of a moth wing or with a breeze" and "climate" has a more detrimental effect, but in this case "one must work with tireless efforts to understand what the rules imposed by nature to itself are" (Festetics 1819c, p. 170).

The debate swiftly diverted after C. C André noted that such "heterogeneous and homogenous nuances" might be hard to notice among plants but it is certainly easy to observe them among humans since "blue-eyed blonds exhibit weaker constitution when several generations marry in the closest possible relationship" (André 1819, p. 26). The question remains, noted André, "how will the position of Nature work in the case of our civilization?" are the similar regularities observed among animals "*eo ipso* detrimental for the full health of the organism" or does inbreeding have deviating influences in humans? (André 1819, p. 26). Festetics's opinion was that although "his natural historical abstraction does not include people" he sees substantial evidence for his thesis to conclude that "Nature in our civilized life does not produce debilitation through inbreeding" (Festetics 1819c, p. 170). However, in the case of a so-called civilized person, the intellect has to be considered alongside physical traits since, in civilization, "scrupulous choices without prejudices, a great education system, and the traits of vivacity, health and mother love" are an important part of upbringing, which constantly alters degeneration (Festetics 1819c, p.170).

It is not Nature that degenerates people in society, because "Nature remains true to its creations"; it is "civilization and domestic discipline (*Huaszucht*) that weakens the primordial force (*Schwächung der Urkraft*)." Festetics was convinced that the "people living in civilization tied by business deals should get closer to a natural way of life, both in their households and treatment of their animals" (Festetics 1819c, p. 170).[28,29] As to C. C. André's question he added that he does see "characteristic facial features, manner and behavior" among "different races" living in small, depopulated areas of Hungary where these *"échappées* [runaways]" intermarry among each other (Festetics 1819a, p. 10). However, he asked "whether among humans inbreeding would be possible by scrupulously observing his points?" (Festetics 1819c, p. 169). In his answer André pointed to aristocratic and royal families who practice consanguineous marriages as an example:

> There are princes and other families, where this closely observed bond indeed expressed a striking *air de famille* [family resemblances], but not directly to the advantage of the descendants, whereon debilitation cannot be overlooked. Perhaps this example explains my meaning better than a long deduction.
>
> **(André 1819, p. 26)**

These were dangerous statements even in a harmless natural historical and animal breeding debate from a person who had previously been characterized with a "French friendly attitude" by the Coachman's secret police. Presumably, these statements had a life-changing impact on André's life and, through it, on the entire long-term functioning of the private learned society of Brno.

Blood and Heredity: Noble × Common

The references that André and before him Festetics made in his statement was a good demonstration of how animal breeders connected heredity with blood. Their best abilities were contested when they faced the danger of degeneration influenced by climate and soil, which were seen to be vying for control. They thought that through their progressive accomplishment the application of controlled mating and selective breeding was advantageous. These methods can be applied on long-term basis to maintain traits in the next generation or to introduce variation. This type of observation argues against a relationship between weather or climate factors being majorly decisive in their approach, since it highlighted a bond between parent and progeny that couldn't be disrupted by other external conditions, regardless of any environmental or climatic influences. This was the main driving force for the survival of a breed (race), which could withstand environmental changes or artificial stimuli preserved in the bloodline, which could yield a successful result if nurtured correctly.

The Central European breeders' view on heredity though blood was an ancient concept dating to the antiquity accepting the tradition of Aristotle. This principle of heredity required blood circulating into the semen, which was regarded as the source of new life. Semen was once thought to be "a foam of the blood emerging from disruptions caused by heat," according to a long tradition first recorded by the ancient Greeks (Brown 1987, p. 184, quoting Diogenes of Appolonia). For decades, there was no rival to the blood/foam idea. According to centuries-old atomists' beliefs, the blood bore in it fragments, originating from the whole body, that were gathered up by the testicles. Galen proposed in Rome that the female produced seminal fluid or liquid from her blood, and Descartes (1596–1650) proposed that the embryo was created in the womb by the fermentation of seminal particles produced by both male and female blood (Pinto-Correia 1997, p. 86). Darwin (1868) defended the term "pangenesis" that described seminal fluids and other liquids circulating from all over the body.

The generation of heat during coitus, referred to as "vital heat" was inextricably connected with the transformation of blood into semen. Foam as a vital quality of sperm was related to poetic representations of Aphrodite, the goddess of procreation, being born from sea foam, which was still being referenced in European literature at the end of the eighteenth century (Roger 1997, p. 41). Thus, the general opinion among nineteenth-century Central European breeders was that there was an unbreakable bond between blood and progeny. The seed gave birth to the blood, which in turn gave birth again to new blood, which was then incorporated into flesh. There were well-known passages in the Bible that affirmed this tradition for anybody who was in question.[30]

Whether or not blood was truly responsible for heredity the language associated with it applied during the inbreeding debates in

Brno offered a widely agreed means of distinguishing degrees of genetic purity, from which degeneration was often a possibility. This proved to be a double-blade sword for breeders. It relentlessly challenged their abilities of controlling mating or crossing animals thus their success strongly depended on managing rearing conditions and selective breeding. The notion that the expression of an animal's inner nature ("blood") could be permanently regulated by human intervention, freeing it from reliance on the natural environment, was a new philosophy that was gradually gaining recognition among Central European breeders. A big problem for sheep breeders was whether a merely "noble" rated stock could be turned into a "pure noble" stock. It seemed that the solution lies in the treatment of the organisms but in what ways? For the early nineteenth-century breeders, the relationship between "nature" and "nurture" was a major concern. Does the blood of the animals evidently get weaker in crosses from "noble" × "common" stock animals? Does this also apply to humans as well and also expressed by André to noble families and prices too?

Gianna Pomata has shown that the discussion of inheritance stemming from recurrent health problems in the eighteenth and early nineteenth century was also a critique aimed at the aristocratic family model (2003, pp. 145–152). Hereditary aristocracy rested on the idea of the transmission of the "noble essence," which required a purity of bloodlines without any interference through interclass marriage (Wilson 2007, p. 143). In the French milieu of the nineteenth century, physicians abandoned the old Galenic traditions and advised the noble families—as part of the practice called *hérédité*—on how to retain the essence of nobility in their marriages to avoid diseases and defects (López-Beltrán 2004).[31] Even the lower classes were aware of the "noble diseases" of the aristocratic and royal families, who were hereditarily predisposed to gout, tisick (tuberculosis), colic, and madness (Figure 2.13). The rapidly growing classes of the landless poor who believed themselves to be the subjects of this hereditary system mocked the "noble features" and "degeneration" of the aristocracy (Cartron 2007, pp. 155–174). To transfer the noble essence of the aristocracy a "noble" × "common" scenario was something that should be avoided.

FIGURE 2.13
The watercolor of Walter Sneyd depicts the satirical features of noble families. Gout was believed to be inherited together with noble titles. (Source: The Wellcome Library No. 44645i.)

Notes

1. Virgil's works had been apparently influencing Hungarian literature and culture since the fifteenth century. In fact, the ideas of antiquity played a decisive role in European science up until the late eighteenth century. As to Hungary, *Georgics* possibly encouraged Count György Festetics to organize his advanced college of farming under the name *Georgikon*, the very first in this kind in continental Europe.
2. Today it is Brno in Czechia.
3. The name Festetics is pronounced as ['feʃtetɪtʃ]. *"Cs"* is a digraph referring to a letter in the extended Hungarian alphabet. In historical documents the form Festetits is also common.
4. Moravian Archive (MZA) in Brno Fascicle 423, Nr. 186; Fasc. 424, Nr. 189; Fasc.. 426, Nr. 2 and 13.
5. Klácel hoped to serve as a philosophy professor in Prague, but due to political persecution, he could only work as a monastery librarian until 1868, when he eventually immigrated to the United

States. He left the Church and went on to work as a freelance writer for journals promoting egalitarian social theories and then finally adopting an idealistic atheist stance. He corresponded with Mendel on a regular basis.

6. Partial holdings of the Austrian Reichstag ÖStA-HHStA-LA-ÖA, 1848, box 90, r. 22-3 (318).

7. Ferenc Széchényi spelt his surname with two diacritics (é); his son, István changed it to Széchenyi, dropping one of the marks.

8. Letters of Hugo Salm to Ferenc Széchényi, Brno, 22 October 1816, MOL P 623. Salm and Hormayr were exchanging letters at least till 1829 about various subjects, including politics and national independence.

9. He was not only in possession of an enlightened library, but his son's tutor Dr. Johann Blaha later sympathized with the French Revolution. Karl Salm supported the French civilization model and had early influences of *Frühliberalismus* in Moravia. His son's scientific aspirations started with alchemic experiments, which were later fueled by the Industrial Revolution in England. Information from a private letter by Jiri Kroupa to Christopher McIntosh from 1989.

10. In 1817, André also built an apparatus to obtain gas from coal; he used the gas to illuminate his coal-heated apartment (Syniawa 2006, p. 27).

11. These constitutions can be found in the Moravian Archive (MZA) in Brno under the designated name of each society, e.g., *K.k. mähr-schles. Gesellschaft zur Beförderung des Ackerbaues* etc.

12. As far as we know these experimenters were only male. One exception we know of is Julianna Festetics the sister of Emmerich and György Festetics who had interest in geology, mineralogy, and botany. She did not become an official member of any *Gesellschaft* but she was involved in shaping the intellectual background for the foundation of Georgikon with her younger brother György. She also made longer trips in the late 1780s and 1790s to Germany, France, Belgium, England, and Scotland with her husband Count Ferenc Széchényi. During these trips she collected minerals and plants. These important collections formed the basis of the establishment of the Hungarian National Museum on September 9, 1803, after her donation.

13. Bornite, also known as peacock ore, is a sulfide mineral with chemical composition Cu_5FeS_4 named in honor of Ignaz von Born.

14. Oddly enough, in this article André misspelt Pictét's name, referring to him as Picket.

15. Located about 60 km from Brno, Hoštice is now part of the Czech Republic.

16. Again, it is odd that André published Petri's articles by having printed the author's name backward, as "Ir lep."
17. Inbreeding is the procedure, used in animal husbandry to date, when animals of the same breed, flock, or family are mated. This is done in order to strengthen traits, found in pure-bred animals, which are seen as favorable. Such a breeding system has the advantage of producing results quickly, while it is disadvantageous because, in the long term, the genotype of progeny weakens due to inbreeding depression. Therefore, successive generations will be more susceptible to certain disorders, and hereditary diseases may also appear in the stock.
18. For "pure animal race," Petri used the term *reine Racethiere*.
19. Names were given by André in publications appearing 1802 onward because of the frequent parallels drawn between the persons. The names also reflect the route of knowledge transfer. After Köcker's study (1809) the names appeared frequently in journals across the Habsburg territories. Unfortunately, not much survived about the activities of Ferdinand Geisslern, only a couple of personal letters; and that he named his famous ram after himself to Ferdinand (Bartosságh 1837, p. 307). His father was a judge in Znojmo, who was awarded with the title of Baron (*Freiherr*) for his services. Geisslern was a soldier who, after his military service, started sheep breeding in Hoštice around 1782. His farm became the Moravian Mecca of sheep breeding, visited by many experts from France, Silesia, Prussia, and Hungary. An Anonymous writer from 1818 mentions that Count Imre Festetics visited Geisslern's estate accompanied by a fellow breeder David Chernel (1790–1808) (Anon 1818a). The name Chernel is presumably misspelled in the source. For further details see André's (1802) report on Geisslern's estate in Hoštice; notes by the local historian d'Elvert (1870) and the summary by Wood and Orel (2001), pp. 191–195.
20. For the original full text, see d'Elvert (1870) II. pp. 145–152, who republished André's version from 1804. A reference to André's original version is found in K in Mähren (1811). The original verse in German is the following: „*Wo frischer Kräuter Balsam düftet;//Der den Schafstall fleißig lüftet//Nicht vom Vorurtheit' beirrt//Wird auch von Müttern, die zu Gatten//Brüder, Söhne, Väter hatten,//G'sund und edle Lämmer zieh'n, -//Die durch deſſen Klugheit blüh'n//welche ich aber vor mehreren Jahren nachfolgender.*
21. Although André rarely signed these articles by name, it is presumed that he wrote them.
22. This journal was conceived with coeditor V. H. Riecke in the spirit of Moravian patriotism and it intended to focus on various cultural and intellectual events in Moravia with the potential of connecting

Czech society, although it was published in German. It was the only and most widely published national newspaper of that time since the period of Joseph II.

23. André wrote the following in the footnote: "This essay was still intended to be published in *Patriotische Tageblatt*. All readers may agree and wish that the incisive and knowledgeable author may continue his views and teachings." The paper was not signed by André but only an abbreviation "A." is found at the end of the paper. Given the context it was presumably written by him. The censorship against André's papers and the suspension of the PTB is detailed in the secret police report AT-OeSTA/AVA Inneres PHSt Z 190 (Austrian State Archives) 1805.

24. The journal took the symbolic name of *Hesperus* "the bringer of light" in reference to *Hesperos* (Έσπερος), "the Evening Star". The name comes from Greek mythology; *Hesperos'* whose twin brother is *Phosphorus* (Φωσφόρος), "the Morning Star". Together they represented the two stages of the planet Venus that is used as a symbol in philosophy and it is a direct reference to the entity of light in which an association to freemasonry and to the Enlightenment is concealed. *Hesperus* was a success with subscribers from across the Habsburg Empire.

25. *Convictus nobilis* was the Latin term for an "education center, boarding school" for young nobles.

26. Imre Festetics jokingly called the new association the "Sheepybunch" or "Sheep-Wise Company of Brno" (Hungarian *Brnoi Juhos Társaság*).

27. Most of the terms used by MAS members in connection with the process of heredity can be traced back to German legal and financial terminology. These terms were used by eighteenth- and nineteenth-century race and breeding studies triggering the construction of the genealogical model of heredity. The financial term *Anlage* meaning "investment" or "trait" was taken up by contemporary thinkers to metaphorically refer to "disposition" or "potential" in a hereditary context as the source of something that needs to be developed. This biological interpretation made it possible to think of dispositions or *Anlagen* of hereditary traits as future potential, e.g., modified by the environment, but linked to an individual from the very beginning in the organization of ancestors. It was Kant who further defined the term biologically in his works *Von den verschiedenen Rassen der Menschen* [Of the different races of humankind] (1755) and *Berlinische Monatsschrift* [Berlin Monthly] (1785) without denoting a specific location of the *Anlagen* within an organism (see Müller-Wille and Rheinberger 2007, pp. 17–18; Lehleiter 2014, pp. 35–45).

28. From antiquity to the nineteenth century, *Hauszucht* or *disciplina domestica* was a means of punishment and education, a measure of chastisement in the Christian family, and a criminal law in feudal rule. It was based on the premise that any kind of domestic discipline must be exercised through physical violence. It was part of a domestic household, but it was also practiced in the form of punishment by landlords on servants. It arose from property rights of landlords, but it was also known in patrimonial court. *Hauszucht* of the landlords for example could be practiced urging the maid-servants to their duties and to punish them when they failed their duties (Anon. 1774, p. 24; Julius 1846, pp. 137–145).

29. By the early 1780s scientists and natural thinkers, e.g., Buffon, Maupertuis, Baumann, Diderot, Robinet, and Bonnet considered humans as the model in the theory of types and they were continuously looking for similarities between humans and various animals. Prominent figures of German *Naturphilosophie* such as Herder and Goethe approached similarities from the opposite direction; concluding that the same force modifying animals' physical formation must be responsible for changes observed among humans. Herder applied this theory to the so-called natural type, where a manifestation of a driving force called the *Urkraft* present within all nature drives animals (*Urtier*) and plants (*Urpfalnzen*) from their archetypes to different physical forms (See Goethe 1790; Herder 2002, 2006; Bentley 2009, p. 205). By using the term *Urkraft*, Festetics must have been aware of these theories.

30. For example, each kind of animal or plant "yields seed or fruit after his kind, whose seed is in itself" (Genesis 1, v.11), that the seed is "conceived in the womb" (Luke 1, v.31; 22, v.21) and the new generation is formed there by God "from the womb" (Isaiah 44, v.2), becoming the "fruit of the womb" (Genesis 30, v.2) and even the (Wisdom of Solomon 7, v. 1,2). The term "seed" was also used in the Bible to refer to an individual's offspring, either male (Genesis 38, v.8) or female (Genesis 38, v.9) (Genesis 3, v.15).

31. The term "*hérédité*" was introduced into English and became "heredity" sometime during the 1860–1870s (see López-Beltrán 1994).

3

From Sheep to Peas

The Transformation of the Sheep Breeders' Society

> The history of a science, art, etc. is often as instructive as science itself. It forces us to compare our present knowledge with that of the past, and since one has to think more in all comparisons than simply looking at it one-sidedly, thus the history of a science often compels us to think more than science itself has taught itself.

> **Johann Karl Nestler (1831)**

In the vibrant intellectual milieu of the Sheep Breeders' Society (SBS), C. C. André could anticipate, already in 1814—when the SBS was founded—that their activities might bring forth some seminal discovery, which was foreshadowed by Imre Festetics in 1819 (Illustration 3.1). However, we have to raise the question as to why their discovery did not lead to a scientific breakthrough in 1819 or later, and why further experiments failed to follow this line of inquiry when SBS members:

1. were properly motivated and committed;
2. had adequate practical background that rested on an adequate scientific basis;
3. supported their practical observations with experiments conducted over a long period of time.

In fact, the success of these experiments would have also generated economic benefits. An unequivocal answer could be to say

DOI: 10.1201/9781003184973-4

ILLUSTRATION 3.1
Imre Festetics makes notes on his Mimush sheep breed.

that SBS members were not interested in heredity because they focused only on animal breeding, particularly the method of inbreeding. Is it possible to deal with animal breeding and its theoretical background without making inquiries into what it is that is inherited over successive generations? André's opening speech delivered at the SBS foundation meeting clearly shows that it was not the case in the beginning; the association had a dual *modus operandi*, being driven by both scientific and economic objectives. By 1819, the structure of SBS had drastically changed, bringing to the fore the "innermost secrets of nature," its "physiological and genetic laws" and their connections, which were not yet examined on a mathematical basis at that time, but some members already had conceptions and plans to do so. It is highly evident that these ideas appeared among SBS members from the mid-1830s. Their breeding experiments and debates in the period of

1816–1820 clearly focused on the timeless issue of heredity. Imre Festetics was well ahead of his age, and arrangements within the association were also utterly unique. No other parts of Europe could witness the emergence of such an intellectual workshop, providing a platform for the reconciliation of so many disciplines and such a great variety of views. To deepen the understanding of SBS activities and their influencing factors, let us examine the historical context in which members of the association pursued these activities.

Academic Freedom in the Age of Metternich

Upon André's arrival to Brno, the Habsburg Empire participated in all alliances organized against the French Revolution and Napoleon. Its aristocracy voted for new taxes, and the army recruited millions of soldiers, with tens of thousands dying in European battlefields. The provision of military supplies created an increased demand for agricultural products. Soldiers needed food, weapons, and clothing, and to produce the latter wool cloth was required as a raw material. Thus, cereal and wool prices began to surge, providing better opportunities for both producers and merchants everywhere in the Monarchy, in Moravian Brno, and in Hungarian Kőszeg too. The Habsburg government had scarce resources to finance its increasing expenditure, thus the Court supported all activities that aimed to produce larger volumes of wool at lower prices. To mitigate surging state debt, they issued bank notes, whose value had to be significantly depreciated in 1811 due to the prolonged war.

By contrast, Moravian and Hungarian landowners enjoyed the favorable period of "quiet rains and long wars," thus the Agricultural Society of Brno could also function without any obstacle up until Napoleon's defeat. As a closure to the Napoleonic Wars, on September 26, 1815, the Holy Alliance was formed by the Russian Tsar Alexander I, the Austrian emperor and Hungarian King Francis I, and Prussia's King Frederick William III—a treaty that set the course of events in European politics until 1848. The Alliance was designed to maintain the feudal monarchy in a Christian spirit, which was envisioned in the framework of a Europe-wide peace-forging and peace-keeping system. Prince Klemens Wenzel Nepomuk Lothar Metternich-Winneburg zu Beilstein (1773–1859), hereafter Klemens von Metternich, the "coachman of Europe," sharply repressed liberal and anti-monarchic aspirations (Proudhon 2011, p. 698). His goal was to eradicate the "French disease" and uncover a supposed "Jacobin conspiracy." Relying on the network of the secret police (*Inneres Polizeihofstelle*), Metternich could, in his own words, "keep an eye on everything. My contacts are such that nothing escapes me" (Sauvigny 1962, p. 105). This special branch of the police was set up to collect information on and prevent political and moral (*sittliche*) crimes and subversive activities pursued by domestic and foreign persons (Hanson 1985, p. 39). On his arrival in Vienna, Ludwig van Beethoven described these conditions as follows:

> Several important persons have been imprisoned here. It is said that a revolution was about to break out, but I believe that as long as the Austrian has some dark beer and little sausages he will not revolt. Briefly, the gates to the suburbs must be closed at ten in the evening. The soldiers have loaded muskets. One doesn't dare raise his voice here, otherwise the police find lodging for you.
>
> **(Prod'homme 1961, p. 18)**

Metternich hired hundreds of informants and snitches so that the secret service could also record the conversations of ordinary people and discover public opinion prevailing within the empire. This "news" was delivered directly to the *Doppelkaiser*, Franz Joseph Karl/Francis Joseph Charles of the Habsburg-Lorraine House, King of Hungary, Bohemia and Germany from 1792, and Holy Roman Emperor as Francis II, finally Emperor of Austria from 1804 as Francis I. Allegedly, the emperor's daily *Morgenplaisir* consisted of hours spent reading through surveillance reports, while Metternich enjoyed dropping tidbits from secret police findings in order to cultivate an aura of omnipotence (Goldstein 1983).

The Empire maintained control over its education system to ensure that "subversive" elements did not infiltrate academic circles and student associations (Goldstein 1983). Teachers "who have obviously demonstrated their incapacity to fill their offices ... by misusing their proper influence on the young, or spreading harmful theories inimical to public order and peace or destructive to existing political institutions" were expelled (McClelland 1980, pp. 219–220). The secret police aimed to discover "dangerous thoughts" and to "nip these in the bud" (Doyle 1978, p. 249), thus another special police department, the *Oberste Polizei- und Censurhofstelle* imposed censorship of newspapers, magazines, and scientific publications in order to uncover a supposedly spreading "Jacobine conspiracy." Newspapers were controlled, magazines and articles had to be authorized, and the texts, books, or other publications considered to be dangerous were confiscated or prohibited, and when necessary, their writers and occasionally mistaken censors were punished too. In this period hundreds of scholars, teachers, and academics were fired for political reasons. Informants and snitches recorded even the titles of books that professors borrowed from libraries. Philosophy as a discipline was barred from study "in view of the scandalous development of this science," whose value "is unproved, but harmful effects from it are possible" (Alston 1969, p. 32). Students were banned from study tours abroad and they had to wear uniforms and have their hair cut in a certain style. They were also barred from visiting coffeehouses or reading newspapers. It may seem ironic that Hungary's 1848 revolution and struggle for freedom set out from such a dangerous coffeehouse.

The Spirit of the Sheep Breeders' Society

Moravian economic development was accompanied by cultural and intellectual advancements, which brought forth the emergence of a vibrant cultural life and the spread of patriotic views. The latter was mainly due to emergent *Vaterlandskunde* movements: various scientific and cultural associations which organized regular (weekly, monthly, or annual) meetings for their members. Brno's Agricultural Society and SBS can be considered such movements, focusing on the study of local features (depending on their seat, Moravian or Hungarian).

Members were equally interested in geology, topography, discovering, and making inventories of natural resources in their country, changes in demography, history, ethnography, and folk art, without explicit political objectives at that time. For example, Count Hugo Salm had a habit of collecting folk songs, but other SBS members were also committed to such ideals, advocating Josephine Enlightenment. They strove to share their ideas with the public through periodicals such as *Hesperus*, edited by C. C. André (Pražák et al. 2003, pp. 50–51). Their ideas reflected the thoughts of Joseph Hormayr (1781–1848), imprisoned in the Spielberg (Špilberk) fortress at that time (Almási 2016). The former "royal seat of Moravian margraves" in the city of Brno served as the prime jail for political convicts. Italian poet Silvio Pellico described his prison years in his book *My Prisons (Le mié prigioni)*, revealing the true nature of the Austrian rule for the broadest strata of Italian society. In 1816, Hormayr was released from the Spielberg dungeons, and Hugo Salm invited him to stay at the castle in Rájec, which also drew the attention of the Viennese secret police.

Both Moravian and Hungarian nobility mostly supported these intellectual movements. They overtly objected to censorship and demanded the freedom of the press; the response to these demands, however, was the imposition of further restrictions. Following the end of Napoleonic Wars, demand for wool cloth used for making military uniforms fell, as did in turn the price of wool. Since there was no need to increase the quality and quantity of wool at any cost and quickly, the Habsburg Empire changed its relation to the SBS. By contrast, breeders insisted on their aim to promote economic, cultural, and intellectual development. Its members, coming from different parts of the Monarchy and from various European countries, shared their ideas through private letters. For instance, the correspondence between Imre Festetics and Hugo Salm covered national, scientific, economic, as well as literary issues.[1] Moreover, SBS members founded similarly organized associations in their home towns, such as the one established in 1815 by Imre Festetics in Kőszeg. The members of these mushrooming "clone societies" attended its meetings or undersigned its documents of association. These gatherings were strongly imbued with the spirit of patriotism, which the Viennese secret police readily noticed and regarded as a threat.

Consequently, they hired informants and *agents provocateurs* to spy on the members of such associations and to watch their activities so that they could nip the spread of the "French friendly attitude" in the bud.

Christian Carl André's Exit

The elder André tried to survive the unfavorable conditions of late feudal absolutism in the age of the Holy Alliance under the cloak of Austrian patriotism. Secretly sympathizing with patriotic movements, Count Lažanský succeeded in protecting C. C. André from censorship and from curtailing his extensive international correspondence, which generated suspicion. However, when the count resigned as Governor of the Province, the pressure of state authorities on SBS members and censorship of their journals increased. Their letters were intercepted by the secret police, which sent reports of them to the Court.[2] These activities were carried out in dedicated mail-opening departments (*Logen*), attached to the post office in every major town and settlement (Á. Deák 2013). Hired informants made copies of the letters, which were then sent to Count Joseph Sedlnitzky (1778–1855), who reviewed and signed the reports supplied by his extensive network of spies and informers (Pajkossy 2006).

These reports designate C. C. André as "a treacherous and useless trickster," a description that "suits his son, Rudolph, too."[3] According to Sedlnitzky's reports, the elder André was driven by the insatiable ambition of becoming a demagogue, a propagator of liberal revolutionary ideas, a characteristic accompanied by morbid vanity (Novotný 2002). Efforts made by André and Salm to establish an institute dedicated to the study of natural sciences proved to be suspicious for the Viennese secret police. However, they successfully asked for financial support from the prestigious local Auersperg and Mittrovský families. They aimed to form an institution in Brno, modeled after Hungary's Georgikon, whose advances were often reported in *ONV*. The Viennese Court did not see this as a desirable development because they thought that it would lead to the strengthening of ties with discredited György

Festetics, who lost even his clerical position at the Chamber. Although György withdrew from public affairs, his brother Imre had significant connections with the society of Brno. The Court could not clearly see what kind of views would guide education or breeding activities at the institution to be established. Therefore, Salm and André envisioned that the institute they were about to found in Moravia could follow the example of the National Museum established in Hungary by Ferenc Széchényi (1754–1820). They asked for Széchényi's advice in several letters.[4] Finally, in July 1817, they founded the institution in Brno under the name Francis (Franzen's) Museum[5] after Emperor Francis I, in order to ward off suspicions.

It was an outstanding event at the 1819 SBS meeting when, in his communication mentioned above, Imre Festetics described heredity as "the genetic laws of nature" in conscious and stark contrast with his opponent, baron Ehrenfels arguing for the "physiological laws" in heredity. The biological significance of his discovery was heralded by the double issue of *ONV*, which also had two separate supplements in the same year to report on the meeting and the importance of inbreeding. They showed the conception outlined in the reports on Festetics's method as innovative, one that is opposed to the religious views of incest. Lehleiter also discusses in detail the questions raised by the SBS as well as the religious aspect of heredity (2014, p. 51). According to her theory, the acceptance of innovative sheep breeding methods would have required the strict separation of religious and biological questions—a step that society could not yet take in 1819, thus animals created through inbreeding were considered monstrosities deriving from forbidden incestuous relationships. Lambs were seen as religious symbols, free from the bounds of matter. However, if animals are formed by biological factors, then they can be transformed by human will, challenging divine creation (Lehleiter 2014). Moreover, Festetics thought that his "genetic laws" apply equally to all living beings, from plants, animals, and humans as well as from nobles to commoners.

The Francis Museum finally opened on May 5, 1820, with the aim that it would promote further studies of issues concerning heredity. In parallel, C. C. André published communications regarding individual rights and liberties in *Hesperus*, which also included the rights and requests of workers employed in agriculture and the wool industry. Employers who controlled agricultural activities represented a conservative stance in relation to inbreeding and wool-working too. In line with the employment morals of this age, children aged 6–12 were purchased from orphanages or workhouses to make them work as doffers in wool works. Doffers were responsible for collecting fallen pieces of cloth from under factory machines and removing rags stuck between moving machine parts. To avoid loss through the complicated stopping and restarting of machinery, small children were seen as fit to climb under the machines. Needless to say, it was extremely dangerous work, which often led to these children's loss of limbs or, at worst, claimed their lives.

André proposed the idea that a commonly comprehensible "National Calendar" should be prepared, which could "contribute to developing broadly interpreted political, arts and technical intelligence in which, based on historical experience, all great advancements are rooted" (Anon. 1821). André had these critical comments, originally published in *Hesperus*, reprinted in the 1820 issues of *ONV*. Presumably, this induced the launch of censorship and political intervention against C. C. André. Secret-service reports do not reveal what the specific reason for this action was, but it seems the most probable from the published journals, since on May 20, 1820, C. C. André resigned from all positions he held in the Brno association, and was followed by several other members. Hugo Salm published a call for SBS members to stand by André and praise his merits, which he also planned to summarize in a petition in order to prevent André's removal from his office. Responding to this call, Imre Festetics also defended André in his letter to Hugo Salm (Appendix 4).[6]

Unfortunately, no support, praise, or paper on behalf of André was sufficient to prevent the axe from finally falling on this excellent organizer, writer, teacher, scientist, and economic adviser (Wood and Orel 2001). According to police reports, André left the city of Brno in September 1821, after he became *persona non grata* for his liberal views in the entire territory of the Monarchy. He moved to Stuttgart, another city with thriving cloth manufacture, where he became an adviser at the court of King Wilhelm I. André's departure from Brno was, at a minimum, a symbol of the restrictions that inhibited intellectual life in this city

(Freudenberger 2003, p. 235). To support this, we can also cite the thoughts expressed by Francis I himself in 1821:

> I have no use for scholars, but only for good citizens. It is up to you to mold our youth in this sense. Who serves me must teach what I order; who cannot do this or comes along with new ideas, can leave or I shall get rid of him.
>
> **(May 1963, p. 79)**

With André's removal, Moravia had lost the leading figure of its agricultural and industrial development, the chief organizer of the SBS. Research in the natural sciences and the questions of heredity lost momentum or even became impossible for some SBS members. The prominent figures of the association and of debates on the study of heredity, Hugo Salm and Imre Festetics,[7] remained silent, while C. C. André left Brno. However, those who had not become targets of secret-service reports, J. K. Nestler and J. M. Ehrenfels, could continue their work.

After André's departure from Brno, his sons Rudolph and Emil remained there. Although André tried to coordinate this work from Stuttgart, managing the affairs of the SBS, physical distance, and secret service surveillance criminalized his work and of those cooperating with him. A confidence gap consequently began to emerge between SBS members over the issues of inbreeding and heredity. As Lehleiter (2014) details it in her excellent monograph, society too was divided over the question of inbreeding, seen by many as "incest." In the SBS debate, the side represented by Festetics, André, and Count Salm held that inbreeding could help create "noble" lines which concentrate useful traits. By contrast, the other side feared the unpredictable and dangerous consequences of inbreeding, particularly degradation in fertility (Wood and Orel 2001). Most SBS members even thought, it a futile attempt to initiate a discussion on inbreeding. Debates on consanguineous mating contradicted. Without André, it proved to be impossible to tip the balance in order to promote the continuation of heredity experiments based on inbreeding methods.

Despite André's attempts to go on with this work, mainly through support from his son Rudolph, the Viennese secret service managed to hinder their activities, preventing communication with the elder André. Imre Festetics wrote in one of his letters that he knew how to contact C. C. André. Initially, SBS members sent their letters addressed to André to Salm's castle in Rájec, who had them delivered to André by way of publisher and editor Johann Friedrich Cotta. However, the ploy did not work because the Viennese secret service intercepted these letters; thus, communication free from censorship became impossible. André went on publishing *Hesperus* in Stuttgart, including manifestos and calls for support from volunteers and advocates. From 1823 onward, these writings were clearly against the Court, express despair, and are imbued with the spirit of liberalism. In his articles published in *Hesperus*, André honestly wrote about accusations against him (C. C. André 1826). The best example of this is his article *"The Price of a Hundred Ducats,"* an excerpt of which should be included at this point:

> Institutions can be fettered by the secret police. From our point of view, this is a very fertile soil in the history of vexing certain states as well as a historical proof of how revolutions come to be. [...] History roars and undoubtedly proves that all traitors, challengers to the throne and fanatics are bound to be punished, for they agreed to dubious legislation, exploiting the trust of many rulers, which further deepened inequalities. Such treachery cannot be left unattended [...] concealed facts (of which there are many in the closets of the judiciary, the police and even private persons) should be made public in order to shed light on the dark machinations of these people, which could start a remarkable process of healing. All who have the opportunity and incentive to uncover harassment and misdeeds should act now.
>
> **(C. C. André 1823)**

Heredity as Mysterious Phenomenon

> It is highly important for refined sheep breeding in general, as well as urgent questions of the time coined by Prof. Nestler: the hereditary ability of noble stock animals.
>
> **(Teindl et al. 1836)**

By 1819, the sheep breeders of Brno accumulated sufficient knowledge to allow Imre Festetics to summarize their practical experience in his so-called "genetic laws of nature." We know very little of the experiments conducted and of the events that occurred in the 1820s, when scientific animal and plant breeding emerged. Johann Karl Nestler (1783–1842), professor of Olmütz (Olomouc) University, appointed in 1823 and a member of the SBS, made an attempt to incorporate this discipline into the curriculum (Illustration 3.2). Unfortunately, Brno's period between 1820 and 1848 remains obscure, and to date we have no knowledge of Imre Festetics's SBS activities until his death in 1847 either. All we can

ILLUSTRATION 3.2
After the departure of Christian Carl André, the experiments on heredity were taken over in Brno by J.K. Nestler and F. Diebl. In their writings, both referred to the inbreeding debates. Nestler was mainly concerned with the breeding of animals, while Diebl was concerned with the production of new plant varieties especially peas, beans, grapes and pears. The two professors worked in close collaboration.

know today is that he withdrew to his estates in Kőszeg and did not attend SBS meetings or write further articles for specialized journals. However, one thing is certain: he officially remained a member of the association until 1847. This period requires a detailed investigation from the perspective of the historical development of genetics, since significant changes occurred in these years, whose exploration can shed light on important connections.

Following André's departure in 1821, SBS members published their findings in a new journal, *Mittheilungen*. The secretarial post of the association passed to J. C. Lauer[8] (1788–1867), who remained in this position until his death. By 1827, young Professor Nestler of Olmütz University had a well-developed curriculum for the subject "Scientific Animal and Plant Breeding," placing emphasis on sheep breeding. Two years later Nestler began to publish his university lectures as a serialized paper (Nestler 1829), and forged a close friendship with Cyrill Franz Napp (1792–1867), the newly appointed abbot of the Augustinian monastery in Brno. In 1826, Napp became a member of the Agricultural Society and showed great interest in the questions of heredity, inspired by the philosophical writings of the founder of his Augustinian order. St. Augustine attempted to place the philosophical ideas of heredity developed by Aristotle and Hippocrates onto religious foundations. In 1827, already a member of the SBS presidency, the abbot invited Franz Diebl (1770–1859), a self-taught expert in plant breeding, to take a professorial position at the Philosophical Institute. These three figures, Nestler, Napp, and Diebl, did a lion's share of on-going animal and plant breeding experiments in this organization at Brno. Unlike Nestler, Diebl preferred plant breeding to animal breeding, and he summarized his conceptions in a five-volume book (Diebl 1835–1841) (Appendix 5). His findings can be summarized in the following major points:

1. The careful selection of parents is a key to obtaining more productive plant varieties.
2. He recognized that *peas* and *beans* are ideal subjects for experimentation in the transmission of productivity and other traits.
3. He believed that a good agriculturist can communicate his ideas and findings in both speech and writing, and make full use of knowledge gained from other, closely related or "sisterly" disciplines.

Lectures given and articles published by Nestler and Diebl—which increasingly focused on studying the theoretical foundations of heredity (*Vererbung*) and its formative "genetic force"—had a great impact among the members of the Agricultural Society. Nestler, according to d'Elvert (1870), collaborated with André early on in the publication of his journals (*PTB, ONV*). Up until 1830, most of his earlier contributions were written anonymously to avoid societal prejudices toward the taboo of inbreeding. Many of the publications in the journal *Oekonomische Neuigkeiten und Verhandlungen* are dealing with selection procedure and heredity, though, we cannot give an exact account for Nestler's contribution neither I was able to identify the authors of these manuscripts but I think it is safe to assume that Nestler contributed to at least some of them.[9]

In 1827, Nestler published a paper in *Mittheilungen* about the horn formation in wild and domestic animals (Nestler 1827). He highlighted the recent development of the English hornless cattle races and suggested the breeding of such cattle in Moravia. In his paper, he noted that he had seen hornless cattle in virtually every village across the country and proposed inbreeding such hornless animals to create pure hornless offspring. He made a reference from the inbreeding practiced by Geisslern. Two years later, Nestler published parts of his lectures that he developed as a Professor of Agricultural Sciences and Natural History at the University of Olomouc (Nestler 1829). The paper entitled: *"The Effect of Generation on the Characteristics of the Progeny"* was an important contribution for the Agricultural Society's animal breeding curriculum (Appendix 6). Nestler touched upon important practical problems of animal breeding outlining the theoretical explanation for artificial selection and its connection to heredity. He briefly mentioned hybridization to explain certain features in plant breeding.

At their 1831 meeting, Baron Ehrenfels, who was also the chief opponent to Imre Festetics in the earlier inbreeding debate, made the following statement:

> The relation between two entities, known as procreation, brings life to the sterile chaos of matter. This genetic force, the start of life, and the creation of lifeless matter all integrate into this mechanism in a sporadic and incremental way, only through climate and nutrition. Nature herself

obeys her institution; an inflexible principal governs in the organization of forms, not Man's desires and demands. [...] Climate, nutrition and generation remain the levers of Nature in the formation of matter. [...] In the interaction of these three potentials, generation, the genetic force, is the most powerful.

(Ehrenfels 1831)

In his argumentation Ehrenfels used the very terms defined by Imre Festetics. At the 1836 meeting, participants articulated additional important questions. The account of this meeting reveals that, according to Nestler's conviction, "the most essential factor in improved sheep breeding is the inheritance capacity (*Vererbungsfähigkeit*) of the noble stock animal." Napp reminded SBS members that, in his opinion, both sexes can pass on the traits of inner and outer organization to progeny, and he also emphasized that inbreeding sheds light on the existence of the heredity phenomenon. In his words:

> [...] heredity of characteristics from the producer *(Erzeuger)* to the produced *(Erzeugten)* consists above all in the mutual elective affinity by kinship *(gegenseitige Wahlverwandtschaft)* of paired animals. Therefore, for each ewe, a ram with corresponding internal and external organization should be chosen. This process deserves to be the subject of serious physiological study.

(Teindl et al. 1836)

At the 1837 SBS meeting, Napp focused on heredity again and made the following very important statement:

> [...] the debate has completely deviated from the proper theme of inheritance capacity. It does not deal with the theory of breeding operations, rather the question is "what is inherited and how?"

(Bartenstein et al. 1837)

In the same year, in response to Napp's comment, Nestler (1837) published his serialized paper under the title "Heredity in Sheep Breeding." In this, he discussed the relationship between nature and breeders in terms of both animal and plant kingdoms, also cited Imre Festetics in his argumentation:

> The terms species and race in the animal kingdom correspond precisely with the terms species and variety in the plant kingdom. Only Nature produces, through forces beyond the hand of Man, under constant environmental conditions, natural species with unquestionable constancy. Man, however, produces in the manner of the forces of Nature, in the reproductive process and formation of organic bodies, modified deviations. From the moment of their origin such deviations have the chance of increasing or disappearing in succeeding generations according to their inheritance.
>
> **(Nestler 1837)**

Nestler tried to highlight that man can also create "modified deviations" from "natural species," but its success depends on the transmission of traits in successive generations. Thus, he pointed out that inheritance happens to a certain degree. Breeders aim to reach the highest certainty, otherwise they only waste their time. Nestler asked SBS members to share their notes on traits transmitted to progeny from pedigree registers, which would allow them to discover certain patterns (Nestler 1837). In response, Baron Bartenstein (1837) defined the breeding ram with constant heredity and with the ability to secure the transfer of its essence to the offspring. In his less than clear explanation, we can see his emphasis on pedigree and production recording and progeny testing. In response, Ehrenfels (1837) pointed out that this transfer can be achieved "through the genetic force" and to investigate the hereditary ability of stock animals *"principles and rules are urgently needed, and also other axillary sciences such as natural history, anatomy and physiology that could translate an unknown truth into a known one."* Nestler (1837) agreed that it was a mistake to think of the issue as theoretical, and reiterated that this is a practical dilemma since solutions can only be derived from experience. Closing thoughts as he added, it seems, that *"I have thrown the seed of the question in the proper soil in which it can gradually develop to the luxuriant fruit of science, if the embryo is cared for."*

Nestler, Napp, and Diebl held that it was important to continue heredity studies based on inbreeding and considered

Bakewell's work exemplary. From 1823 up until his death in 1841, Nestler had been keeping this issue on the agenda and had tried to convince SBS members for 18 years that they should initiate further debates on inbreeding. The three leading figures attached the adjective "scientific" to the method of inbreeding and thought that the transmission of traits can be unequivocally clarified if these traits are studied in progeny and stringently selected.

According to the notes of MAS meetings, members were not inclined to embrace this idea, some even thought that inbreeding was not worth discussing (Wood and Orel 2001). However, when Abbot Napp put forward the question "what is inherited and how?" the resentful and judgmental attitudes of SBS members began to change. On the twentieth anniversary of publishing Imre Festetics's article on inbreeding, Nestler's (1839) detailed study appeared in the journal *Mittheilungen* under the title *"Über Inzucht"* ("On Inbreeding") referring to the Hungarian count's earlier paper. Besides this allusion to the beginning of an important debate, he gave a thorough analysis of Bakewellian methods and also discussed Festetics's contribution:

> Inbreeding became known by way of Count Emmerich Festetics's work, who, in 1803, created a small closed stock with one ram and nine ewes, where the father had been the son over succeeding generations. His report can be read in number 22 volume 1819 [of *ONV*]. In debating the issue of inbreeding with an advocate of the crossing system, Baron Ehrenfels, the noble count also noted that inbreeding cannot be applied unconditionally.

(Nestler 1839)

After Nestler's death, from 1841 onward, Napp and Diebl undertook to go on with the study of the question concerning inheritance through inbreeding. Since Diebl was a plant breeder, he suggested that crop plants, fruits, vines, and forestry trees, and particularly peas and beans, should be studied rather than sheep (Diebl 1844). He published a book in 1844 on his plant breeding techniques and held related lectures in Brno. These lectures were also frequented by a curious friar called Gregor Mendel (Illustration 3.3).

ILLUSTRATION 3.3
The young Mendel arrived in Brno in 1846, where he became a student of the elderly Professor Diebl, with the support of Abbot C.F. Napp. Mendel obtained excellent examination results in all natural science subjects. Napp and Diebl had an interest and long experience in the study of biological inheritance. Napp believed that heredity was not a practical but a theoretical question; he formulated the research question for Mendel, what is hereditary and how?

Mendel's Peas

The young Mendel could build his own experimental design concerning the wealth of experience in animal and plant breeding that had been accumulating for 50 years in Brno. His experiments aimed to scrutinize "the striking regularity with which the same hybrid forms always recurred when fertilization happened

ILLUSTRATION 3.4
Napp financed Mendel's university studies in Vienna, and after his return to Brno, he had a greenhouse built for his pea experiments, which was extended twice.

between like species (*Arten*)," a phenomenon previously noted by "numerous careful observers" (Mendel 1866a).[10] To conduct his experiments, Mendel chose inherently "inbred," self-fertilizing garden peas (*Pisum sativum*). Inspired by Napp and Diebl, he began to examine variation in plants, seeking an exact answer to Napp's question "what is inherited and how?" through these experiments (Illustration 3.4). He was interested in nature and its diversity, rather than breeding efforts (Figure 3.1). Thus, in 1846, he joined the Naturalists' Society, newly founded as a distinct section within the Agricultural Society in Brno, similar to the SBS. The members of this new association were primarily interested in how species acquire their characteristics and what it is that shapes them. In 1861, they split from the Agricultural Society to form an independent organization. When working with crosses, Mendel carefully removed the reproductive organs of individual pea plants in order to conduct controlled crossings. He made detailed notes

FIGURE 3.1
Members of St. Thomas's Abbey in Brno about 1862. Gregor Mendel is standing second from right, while Cyrill Napp is seated second from right. (Source: Photo courtesy by Jiří Sekerák from the archive of Mendelianum, Moravian Museum, Brno, Czechia.)

of characteristics that appeared in the progeny (e.g., seed shape and color: round or angular, yellow or green) and, after years of crossing experiments, tried to interpret his findings on the basis of mathematical methods, which were considered important by Festetics (1820) too. In this, he could rely on his former training in physics and mathematics at the University of Vienna. He thought that specific characteristics, such as seed color, are determined by distinct "factors" (now called genes), which are transmitted in a rather consistent ratio. Today his insights are referred to as "Mendelian rules/laws." In short, these rules state that specific hereditary factors combine and are transmitted to progeny independently of one another. There are two distinct types of factors. One is masked or recessive and the other is dominant. Recessive characters do not always appear after crossings in the first filial generation, but they can reappear later, in the second filial generation with certain frequency. This happens when two recessive factors are paired. It is essentially a mathematical proof of point "b" listed by Festetics (1819c) under the heading "genetic laws."[11] Moreover, Mendel observed that in hybrid generations there can be individuals with mixed characters, which differ from the traits of both parents.

Mendel also tried to test his findings from *Pisum sativum* experiments with bees (Vecerek 1965), but the transmission of their traits was much more complex, thus he could not reproduce the results observed in peas. Quite disappointed Mendel abandoned crossing and continued his meteorological work instead.

If we looked back at his scientific work, ignoring his earlier achievements, we could certainly consider him a meteorologist because he had many more publications on this subject (altogether eight papers: Mendel 1863, 1864, 1865, 1866b, 1869, 1871, 1879, and 1882). In fact, he published his meteorological studies three years before he commenced his now highly regarded experiments with peas.

Like the "genetic laws" articulated by Imre Festetics, Mendel's rules were ignored due to the lack of understanding during his lifetime. His apicultural experiments, which failed to bring forth reassuring results, confused him, and he died skeptical of his own achievements.

The groundbreaking nature of his work and the significance of the 3:1 segregation ratio[12] went unnoticed.

Bateson's Chickens

Mendel's work attracted little attention, although it was published in the Proceedings of the Natural History Society of Brno (*Verhandlungen des naturforschenden Vereines in Brno*), a specialized journal circulated throughout European private and "public" libraries in numerous copies. It was acknowledged only later, in 1900, during a series of "rediscoveries" when three European botanists (Carl Correns, Erich von Tschermak-Seysenegg, and Hugo de Vries) "independently" proved Mendel's conclusions through their own experiments. English biologist William Bateson (1861–1926) is attributed a significant role in this story because he became the leading proponent of Mendel's work at the same time and coined the term "genetics" in 1905, opening the century of this new discipline. However, such statements pose several problems which need clarification, as highlighted by Szabó T. and Poczai (2019).

In 1866, Mendel sent his findings to one of the most prominent botanists of his age, Carl von Nägeli (1817–1891), then working in Munich. Besides his experiments with peas, Mendel also reported his crossing experiments on hawkweeds (*Hieracium*) to Nägeli. A detailed and prolonged correspondence began between the two researchers, and Nägeli finally dismissed Mendel's work as insignificant, to the latter's great disappointment. Nevertheless, Nägeli informed his disciple, Carl Correns (1864–1933)—who conducted heredity experiments unaware of the findings from the Brno study on peas—about Mendel's *Hieracium* experiments. Correns finally published his results in 1900, with reference to and appreciation of Mendel's achievements (Correns 1900). It is important to note that Mendel never used the term "law" to refer to his own conclusions. It was Correns who summarized them as the "rules" (*Regeln*, not *Gesetze* i.e. laws) of (1) segregation, (2) independent assortment, and (3) dominance. Due to misinterpreting Correns's German term, they were first described in English as "laws" by American biologist Thomas Hunt Morgan (1866–1945), who won the 1933 Nobel Prize for discovering the linkage between genes and chromosomes in fruit flies (Morgan 1919). Meanwhile, Correns married Nägeli's niece, inheriting the correspondence between Mendel and Nägeli, which he later edited and had published.

The Dutchman Hugo de Vries also independently reported his crossing results in 1900 (without even a hint of Mendel)—in the same year as Correns published his findings. He claimed that he reached conclusions similar to those of Mendel, without, however, being aware of the friar's work (de Vries 1900). One of his disciples, Stomps (1954), revealed the truth, in stating that de Vries must have known and read Mendel's 1866 paper. Martinus Beijerinck (1851–1931) sent a letter, together with a copy of Mendel's work, to de Vries well before the Dutch botanist prepared his own paper. Beijerinck wrote the following comment on the attached study to de Vries: "I know that you are studying hybrids, so perhaps the enclosed reprint of the year 1865 by a certain Mendel, which I happen to possess, is still of some interest to you."

The history of an "independent rediscovery" by a third researcher, Austrian botanist Erich von Tschermak-Seysenegg (1871–1962) was recently uncovered in detail by Simunek et al. (2011). According to their chronology, in 1898, Tschermak-Seysenegg went to the Belgian city of Ghent to study botany. On the advice of Professor Julius MacLeod (1857–1919), he started experimenting with self- and cross-pollination in plants, choosing several varieties of peas for his hybridization studies. In the same year he left for Amsterdam to visit Hugo de Vries, with whom he had a shared research interest and a friendly correspondence. After his return to Vienna in October 1899, he began to analyze the results of his experiments in the F_2 generation, which he needed for his inaugural dissertation. According to his memoirs, it was in Vienna, in Adolf von Liebenberg's private library, that he first encountered Mendel's 1866 work, which he was studying between October 10 and December 1899 (Tschermak-Seysenegg 1958). He delivered his thesis on January 17, 1900, and on March 26, 1900 he received de Vries's paper, written in French and German, "The Law of Segregation of Hybrids" (*Sur la loi des disjonction des hybrides/Das Spaltungsgesetz der Bastarde*).[13]

In March 1900, de Vries submitted the German version for publication in the journal *Berichte der Deutschen Botanischen Gesellschaft*. At this point, Tschermak-Seysenegg decided to publish his own results. In May 1900, he received a reprint of Correns's paper (1900), which was published in April 1900, also in *Berichte*, under the title "Gregor Mendel's Rule Concerning the Behaviour of Progeny of Varietal Hybrids" (*Gregor Mendels Regel über das Verhalten der Nachkommenschaft der Bastarde*). At the advice of his brother Armin, who advocated the idea of tripartite discovery, on May 16, 1900, Tschermak-Seysenegg prepared a short abstract of his thesis and sent it to the *Berichte*, the same journal where the papers by de Vries and Correns appeared, and it was finally published on June 2, 1900 (Tschermak-Seysenegg 1900). Today these events of the year 1900 are taught at schools briefly as the "rediscovery" of Mendel's work, which is attributed to Correns, de Vries, and Tschermak-Seysenegg without much explanation. The current syllabus of teaching genetics usually ignores events in Brno before Mendel and the members of the SBS, such as Imre Festetics.

Then there is another peculiar fact, which is the subject of a debate on the origin of the name for the new discipline of genetics. It is justifiable to ask why this branch of science, so significant today, is called "genetics." The most common answer is that it was a certain William Bateson who, in a letter to Adam Sedgwick[14] (1854–1913), first used the term "genetics" in 1905.

I cite below the relevant part from this letter, dated April 18, 1905:

> *Dear Sedgewick,*
>
> *If the Quick Fund were used for the foundation of a Professorship relating to Heredity and Variation the best title would, I think, be 'The Quick Professorship of the study of Heredity'. No single word in common use quite gives this meaning. Such a word is badly wanted, and if it were desirable to coin one, 'Genetics' might do. [...]*

From the 1880s onward, young Bateson showed an interest in heredity and the natural variation of species. He had been fascinated by plants from the very beginning, but experiments in animals were closer to his heart as a zoologist. According to his idea, there must have been a law which had similar impact on all living beings, thus it equally applied to plants, animals, and all creatures living on the Earth. To prove his thesis, Bateson carried out experiments together with his friend and colleague Reginald Crundall Punnett (1875–1967). Bateson's research was derided by the University of Cambridge, thus he had no chance to obtain any scientific grant or job. He worked at the university canteen to sustain his family.[15] In the meantime, during his collaborative experiments with his friend Punnett, they observed that the progeny of crossed black chickens were not always poultry of black feather, some of them were white. They were astonished by these results and began to think how it was possible that two black parents produced white offspring. The more times they repeated the crosses, the more intriguing insights they reached. In all generations of progeny the ratio of black to white chicken was always 3:1. According to their theory, parents passed on some kind of instructions to their progeny, which held the proportions of black and white chickens constant.

Today these instructions are called genes. Only through understanding these proportions, we can recognize deeper patterns of inheritance. Bateson and Punnett assumed that the regularities observed in chickens work similarly in other living beings. Therefore, they started to experiment with dozens of domestic animals in order to prove the 3:1 ratio. However, they had to look after plenty of these animals, which incurred increasing costs and required more work. Due to the scarcity of resources, they had to look for volunteers to help. They recruited assistants from among the students of Newnham College, Cambridge. Since the college had only female students, the volunteers who participated in crossing domestic animals are called "Bateson's ladies" to date.

Later Bateson and Punnett also observed segregation ratios in the transmission of other characters, such as comb shape. For example, when crossing chickens with rose- and pea-shaped combs, offspring grew a new phenotype, walnut-shaped comb. Among the progeny of this first generation, crossing yielded a 9:3:3:1 ratio, with nine walnut combs to three rose combs to three pea combs to one single comb (Bateson 1902a; Bateson and Punnett 1905–1908).

According to Robert Olby (1987), it has long been accepted that the first account of Mendel's work in English was given by the Cambridge zoologist, William Bateson, to an audience of Fellows of the Royal Horticultural Society in London on May 8, 1900. This is based on two sources: the paper "Problems of Heredity as a Subject for Horticultural Investigation," published in the society's journal later that year and stated as "Read 8 May, 1900," and Beatrice Bateson's account of the event over a quarter of a century later. Of the paper which her husband gave on that occasion she wrote:

> He had already prepared this paper, but in the train on his way to town to deliver it, he read Mendel's actual paper on peas for the first time. As a lecturer he was always cautious, suggesting rather than affirming his own convictions. So ready was he however for the simple Mendelian law that he at once incorporated it into his lecture.
>
> **(Olby 1987)**

This was surely the very reason why Bateson and his wife visited the "genius loci" where the 3:1 ratio of segregation had already been recognized 34 years earlier by a certain Gregor Mendel, only he experimented with peas rather than chickens.[16] When Bateson returned home, he had Mendel's work translated into English (Bateson 1902b), and became a proponent of Mendelian inheritance theses.

If he was so surprised to read Mendel's paper, we can only wonder how the British scientist would have responded if he had gotten hold of Imre Festetics's article describing "genetic laws."

Notes

1. Imre Festetics to Hugo Salm, 23 and 27 January 1821, Moravian Archive, Brno G150/K80 and K81.
2. Österreichisches Staatsarchiv (Austrian State Archives), AT-OeStA/AVA Inneres PHSt 22 b; AT-OeStA/AVA Inneres PHSt Z 190; AT-OeStA/AVA Inneres PHSt 2118.
3. J. Sedlnitzky's report on C. C. André, 1823 (Austrian State Archives, AT-OeStA/AVA Inneres PHSt 1823-9239/1065/23).
4. H. F. Salm to Ferenc Széchényi, October 22, 1816, Brno (National Archives of Hungary, MOL P 623).
5. This institution later became today's Moravian Museum (Moravské zemské muzeum).
6. Original in German, translated to Hungarian by István Bariska.
7. Österreichische Staatsarchiv, PHSt ZI.0385–1820; PHSt ZI.6104–1822 fol.01–48r.
8. Lauer still held this position when Gregor Mendel first read his later famous paper in 1865.
9. At the same time, Nestler also collaborated with his friend Karl Jurende (1780–1842) in publishing another newspaper entitled *"Moravia"* often anonymously. Jurende was a strong supporter of C. C. André and respected him as a journalist. Jurende was interested in creating new fruit tree verities and served as the director of the Philantropinum in Kunín between 1806–1810. The same position was held by his predecessor J. Schreiber who became a vicar in Mendel's hometown Hynčice.
10. Mendel read his paper "Experiments on Plant Hybrids" (*Versuche über Pflanzen-Hybriden*) first in 1865, at a meeting of the Brno Natural History Society, which SBS members also attended, with old Lauer presiding over the session. His study was published in the Society's journal a year later only.
11. Traits of grandparents not reproduced in their progeny may reappear again in successive generations.
12. In his experiments with peas Mendel used pure lines homozygous for one or more characteristic, e.g., for yellow (AA) and green (aa) seed color. He observed that all individuals in the first filial generation (F_1) had the same (uniform) yellow seed color (AA or Aa). By contrast, in the second generation (F_2) seed color characters separate in the proportion of 3:1. The phenotype (seed color) of offspring can always be traced back to the ratio of 3 yellows (AA and Aa) and 1 green (aa). Therefore, the characters of grandparents reappear in the F_2 generation. Imre Festetics recognized the same phenomenon, but without a mathematical model, that is, the 3:1

segregation ratio. From this ratio Mendel concluded that seed color was determined by two factors—or, as we call them today, alleles of a gene. Reproductive cells (gametes) carry only one of these factors (A = yellow and a = green), which combine by chance, producing a dominant homozygous (AA) and two heterozygous (Aa) individuals for the dominant allele and one homozygous (aa) plant for the recessive allele. According to the segregation ratio, three individuals carry the dominant allele (yellow phenotype) and the fourth carries only the recessive allele (green phenotype), while the segregation ratio of these genotypes is always 1 (AA) : 2 (Aa) : 1 (aa) = 3 dominant : 1 recessive phenotype.

13. N.B. that Tschermak-Seysenegg investigates the segregation ration of so-called "bastards" (*Bastarde*), the same phenomenon was debated by Ehrenfels and Festetics in Brno using the same word for progeny resulting from cross- and inbreeding, and such "bastards" were subjected to artificial-selection. The question coined in 1819 related to the nature of such creatures.

14. Despite having the same name as his great-grandfather, Darwin's professor of zoology, the two persons, although they are linked by kinship, are not identical.

15. I owe this information to Timothy Walker the *Horti Praefectus* of the University of Oxford Botanical Garden.

16. William Bateson and his wife Beatrice took a holiday based on a sudden decision around December 11, 1904. His visit is recorded in a small booklet from January 5, 1905, and his signature is found in the visitor's book of the Königskloster (December 29, 1904). There is also a personal letter from Ferdinand Schindler from December 13 that arranges a meeting with Bateson. I thank Donald R. Forsdyke for bringing this information to my attention.

4

The Legacy of Imre Festetics

The Festetics Oeuvre

First, I believe it is important to publish a comprehensive list of all works authored by Imre Festetics. Apparently, numerous flaws and inaccuracies have developed in both Hungarian and international related literature with respect to quotations from and content of these works. These must be corrected to make further effective inquiries and research possible. Currently, Festetics's nine papers are known, which were all published in the economic journal *Oekonomische Neuigkeiten und Verhandlungen (ONV)*. In addition, one of his communications was published in an 1815 issue of a Hungarian weekly *Nemzeti Gazda* [National Farmer] (Festetics 1815a).

These publications constitute Imre Festetics's scientific legacy. It is important to note that these works have been evaluated from the perspective of science history only in part, if at all. Of these papers, six discuss the general methods and technical description of sheep breeding. These communications should be thoroughly examined and fitted into the historical flow of sheep breeding debates, so that we could have a more refined picture of the antecedents to the debate between Ehrenfels and Festetics, as well as insight into the breeding activities and methods applied by the Hungarian count in Kőszeg. Inaccuracies in literature have also emerged with respect to the citation of these articles, thus I feel obliged to list their precise forms of reference below:

1. Festetics, E. (1815b) Aus einem Schreiben des Herrn Grafen Emmerich Festetics zu Güns in Ungarn [From a

Paper by Count Imre Festetics of Kőszeg]. *Oekonomische Neuigkeiten und Verhandlungen* 69:547–548.

2. Festetics, E. (1819a) Erklärung des Herrn Grafen Emmerich von Festetics [An Explanation by Count Imre Festetics]. (Vergleichen Nr. 38., 39. u. 55., 1818). *Oekonomische Neuigkeiten und Verhandlungen* 2:9–12.

3. Festetics, E. (1819b) Erklärung des Herrn Grafen Emmerich von Festetics [An Explanation by Count Imre Festetics]. *Oekonomische Neuigkeiten und Verhandlungen* 3:18–20.

4. Festetics E. (1819c) Weitere Erklärung des Herrn Grafen Emmerich Festetics über Inzucht. *Oekonomische Neuigkeiten und Verhandlungen* 22:169–170.

5. Festetics, E. (1820a) Bericht des Herrn Emmerich Festetics als Repräsentanten des Schafzüchter-Vereins im Eisenburger Comitate [A Report by Count Imre Festetics on Behalf of the Sheep Breeders' Association of Vas County]. *Oekonomische Neuigkeiten und Verhandlungen* 4(19):25–27.

6. Festetics, E. (1820b) Äuserung des Herrn Grafen Festetics [A Statement by Count Imre Festetics]. *Oekonomische Neuigkeiten und Verhandlungen* 15(20):115–119.

7. Festetics, E. (1820c) Bericht des Herrn Grafen Emmerich Festetics über die vom Schafzüchter-Verein angeregten Versuche der Heckeisfütterung nach Petrische Methode [A Report by Count Imre Festetics on Experimenting with Petri's Method of Chopped Fodder, Proposed by the Sheep Breeders' Society]. *Oekonomische Neuigkeiten und Verhandlungen* 25(20):193–195.

8. Festetics, E. (1822) Über einen Aufsatz des Hrn. I. R. in 3ten Hefte des Jahrganges 1821 [On an Essay by Mr. I. R., published in the 3rd booklet of the *ONV* volume 1821]. *Oekonomische Neuigkeiten und Verhandlungen* 92:729–731.

For communications that were published in 1820, volume numbers should also be indicated, since *ONV* appeared twice with double issues in that year, each of which was complete with a separate supplement (*Beilage*). These issues and their supplements had continuous page numbering, thus only indicating page numbers can refer to four different volumes of the specific year.

A frequent mistake in literature concerns Imre Festetics's much-quoted paper discussing "the genetic laws of nature," which is listed above as item IV with the exact original title "Further Explanations on Inbreeding by Count Imre Festetics." The correct reference is thus "On Inbreeding" (*über Inzucht*), which appeared in print as the title in the second line.

The count undersigned these articles as *Emmerich* Festetics, using his Christian name in its German form. In addition to his paper on inbreeding, listed under point V, article number VII is also of great significance because it discusses measuring the physical properties of wool, their tracing in successive generations and their mathematical analysis. The importance of mathematical analyses and relations is also demonstrated through practical observations. This article is often confused with Festetics's other remark concerning the application of mathematical standards:

> It will be judged as marking the beginning of an epoch in the science of breeding that in 1819 grades of wool fineness were established and defined with mathematical precision.
>
> **(Salm 1820)**

It is true that these were Imre Festetics's words—which he had never published in an article. In his report of the 1819 Sheep Breeders' Society (SBS) meeting, Hugo Salm quoted Festetics's sentence literally. This occurred frequently on the pages of *ONV*, thus an in-depth investigation may reveal additional notes, quotations, or even articles that were forgotten or written anonymously. Under André's editorship, the names of authors were frequently printed as palindromes (e.g., "Irtep" for Petri) or as complex riddles. From the 1820s onward, the main SBS journal was *Mittheilungen*, which may also contain further quotations from Imre Festetics. There are a number of papers, which were published anonymously about inbreeding after André's departure majorly between 1822–1829 in *Mittheilungen* and also in *ONV*, edited by Rudolph André. For example, an anonymous article from 1824 discusses constancy of traits in subsequent generations of animals, where the author states, that "*Nature teaches from the blade of grass to the cedar, from the mite to the elephant, that in the superabundance of her products, imperfect products, imperfect formations never disappear completely* (Anon. 1824)." The authors close his writing by saying "*Happy are those, who do not see, and yet believe!*" On another anonymous paper from 1827, related to the same subject B. Petri inserted the following footnote "*If we are to take instruction from those familiar with this topic, the species of our systems are not true species of creation, but for the most part new races of our domestic animals, Natural Science has begun a new epoch* (Petri 1827)." These papers are aviating further historical analyses, though we may never decipher the name of the exact authors.

What Did Festetics Discover?

Undoubtedly, Festetics derived his "laws" from his breeding practice, which was primarily aimed at producing wool of better quality. Although initially he did not strive to evaluate his experiences mathematically, his later writings reveal that he was aware of its importance (Festetics 1820a).

He made his groundbreaking observations in an age when "heredity" had no biological significance; in fact, biology itself was a discipline in the making. The thinkers of this era could not clearly distinguish the concepts of inheritance and development. The laws formulated by Festetics reflect that he was very close to making a distinction between inheritance *sensu stricto* and development *sensu lato*. Unfortunately, he chose traits to be studied such as wool density, crimp, and length, which were genetically complex and subject to polygenic inheritance. Festetics would have not been able, simply by recording wool traits in successive sheep generations, to observe the 3:1 segregation ratio, as Mendel did with peas or Bateson with chickens. He would have succeeded in fully quantifying his findings and reaching conclusions similar to those of Mendel only if he had chosen traits of monogenic inheritance.

Festetics also lacked precise techniques and modern statistical methods such as quantitative trait loci (QTL) mapping (e.g., Purvis and Franklin 2005; Bidinost et al. 2008; Wang et al. 2014), which would have allowed him to evaluate the transmission of wool traits, since it was impossible with nineteenth-century methods. This was achieved much later by Sewall Wright (1889–1988), who had been working painstakingly for 30 years to finally lay the foundations of quantitative population genetics. The significance of the main points formulated by Festetics (1819c) was summarized by Szabó T. (2009) as follows:

1. The transmission of traits is governed by internal factors; it is only influenced by external ones.
2. Inbreeding strengthens, rather than weakens, the potential of trait transmission, provided that desirable traits are controlled through rigorous selection.
3. All living beings, including humans, are continuously shaped by the interaction between genetic laws and selection processes.

It is important to note that, although it had not been discovered before Mendel, Imre Festetics laid the foundations of factorial genetics. This work, the "genius loci," is what lead to Gregor Mendel. While Festetics's work (as later Mendel's) did not induce a scientific revolution in his day, his insights led to a radical reinterpretation of ideas concerning heredity. In the course of conducting practical crossing experiments, he reached theoretical conclusions that have remained relevant to date (and are now called "dominance" and "segregation"), recognizing the essential difference between physiological and genetic research. He proved that endogenous factors have a definitive role in heredity and he was the first in the history of science to use the term "genetic laws" in its current sense. From a philosophical perspective, it seems to be the most remarkable feature of his oeuvre that he recognized the primacy of genetic effects (information) over environmental effects (circulation of matter and energy). This insight has truly philosophical aspects.

In the following quotation, Elemér Hankiss (2017) ponders upon such a synthesis:

> There are philosophers who argue that a possible Theory of Everything must reconcile, unify, comprise not only the laws of the theory of general relativity, the theory of gravity, and the laws of quantum mechanics but also those of the realm of human (or cosmic) consciousness, i.e., an ultimate equation which the human mind can understand and

handle. This would make an extremely strong link between
the quantum universe and the human mind, even if it did
not mean that individual human lives have meaning.

The nineteenth century was an age of great economic, political,
social, and scientific transformations. A modernizing European
population demanded increasingly more freedom, inevitably
imbued by the ideas of natural philosophy. This also had an enormous impact on the sheep breeders of Brno and the members of
related societies—organized in a similar manner as the several
sheep breeders' associations of the region, driven by patriotism,
the love of their homelands, and a new approach to nature—and
to society as well. This emerging new spirit, which certainly
had for-profit economic aspects, encouraged breeders to initiate
debates on "the innermost secrets of nature": on the transmission
of traits across successive generations. Scientific questions raised
in relation to heredity very soon became intertwined with politics and philosophy, when "the genetic laws of nature" demanded
equality at birth. To recognize and make public, the fact that all
living creatures of the world, including aristocrats, are subject to
the same rules of heredity was to induce fear in the absolutist
power.

In 1819, when Festetics submitted his work for publication, many
people were unprepared to separate religious beliefs from natural
phenomena, and the same was true for the state of society 46 years
later, when Mendel conducted his own experiments with similar
but more accurate methods. Such a change and acknowledgment
could take place only in the early twentieth century as a result of
simultaneous "rediscoveries." The society of the early nineteenth
century was appalled to see that open-minded breeders tried to
utilize their findings in practice right away in order to produce
wool of improved quality through inbreeding. Consanguineous
matching as an "incestuous" method was seen as artificial modification of nature, which contradicted the cultural and religious
norms of the age. To date, this stigma has haunted our ideas of
heredity and genetics throughout Europe. Attempts at explaining
Imre Festetics's thoughts from this perspective require further
research.

Nineteenth-century society viewed the individual as an entity
determined by inheritance and birth. The study of heredity or,

given a new name, genetics, focusing on the issues of inbreeding viewed as incest and of crossing experiments, did not question the freedom and equality of the individual; on the contrary, it was the direct consequence of these ideas. It was in the early 1800s that the question of abolishing legally interpreted inheritance and related privileges, after a long period of their prevalence, had emerged in social thinking. This raised the issue that individual freedom is not predetermined and hereditary. The transforming image of the individual also provided increasing independence and individualism which, as reflected in C. C. André's writings, can easily generate claims to autonomy. All this happened with the leadership of patriotic movements where the desire for freedom and the independence of Moravia or Hungary was very strong and later, in 1848, led to the Hungarian Revolution and War of Independence. At first sight, linking the questions of nature and heredity with social changes may seem anachronistic, but I am convinced that, due to changing ideas of nature, reproduction, and inheritance, the issues of freedom and autonomy triggered long-term shifts in social arrangements.

Procreation (*Zeugung*) and heredity were once misunderstood and thought to be a jigsaw puzzle in the early nineteenth century. In this regard, Brno set a significant milestone in 1819 by bringing together scientific ideas, empirical observations, and legal concepts relating to heredity from various areas of human experience in order to solve the enigma of inbreeding. The fact that such different currents of thought led to the foundation of hereditary thinking before Mendel in the same city and educational institution where he later worked sets a great prelude for his ideas. It is not far fetching to state that Mendel's laws of hybridization and the characteristics of his novel experiments leading to his discovery were foreshadowed as early as 1819. The activates of Central European breeding societies created an appropriate scientific milieu, setting up a chain reaction in the development of expertise in the topic of hereditary studies, eventually supporting the formulation of main questions of Mendel's immediate research.

Simple features of animal improvement by artificial selection were drawn from adopted breeding experience in Festetics scientific explanations. He also elaborates his arguments from a natural historical perspective by citing references from the animal and plant kingdoms, as well as humans, though these theories are far from systematic and remain at an empirical level in the topic of heredity. Festetics is inspired by the spirit of the Industrial Revolution and the mechanistic thought of the time. One clear example of this is where he states that since the "manifold architecture of the horse machine" is difficult to understand, inbreeding horses are more difficult than inbreeding other animals (Festetics 1819c). In this way, heredity for Festetics tends to be about both scientific knowledge and technical exploitation of generation. These viewpoints are still important today in terms of genetic engineering and looking back in history to obtain a greater understanding of the early convergence of science and technology could expand debates on current topics.

Imre Festetics provided crucial advice to Central European animal breeders, who realized that the traits of parents have an effect on the next generation, and that unintended differences can be long-lasting and persist in the offspring. These attributes can be consolidated in each "organism" by picking certain variations. Thus, human interference can modify trait transmission by artificial selection, in which the breeder assumes a role equivalent to "natural forces" (Figure 4.1). *"From the seed's grain, produced by such a refined artificial fertilization, a new offspring emerges consisting of the characteristics of the father and mother plant"* (Hempel 1820; Appendix 7), writes André's friend G. C. L. Hempel, secretary of the Pomological Association. At the very end of his writing, he mentions peas in a brief remark, which Diebl later embraces and often mentions as an example in his lectures. Hempel also notes that in order to understand precisely and to provide further explanations about experiments with artificial fertilization, a person is needed who has solid botanical expertise, a keen observational mind, and a tireless relentless patience. This description seems to fit perfectly with Mendel, who later experimented in Brno.

According to Festetics, culling should be used to prevent the deterioration of the living organism that occurs during inbreeding:

> Experiments founded on concepts that go beyond the laws of nature or seek to act against it through force, such as Bakewell's, Buffon's, and Sebright's ambitions that go beyond the laws of nature or tried to act against it by force, are less deserving of gratitude; as we all know, striving against natural forces is often met with retaliation through nature. As a consequence, I assume, together with the founders of the

FIGURE 4.1

Correa spp. (Rutaceae) hybrids are widely grown in Australian gardens. Many hybrid cultivars were developed in through artificial fertilization mentioned by Hempel in his paper till the mid-nineteenth century as shown on this colored lithograph from c. 1856. (Source: The Wellcome Collection.)

Agricultural Society and the advice of the widely esteemed Thaer, that conception within the nearest bloodline is not only not dangerous, but also helpful if the aim is to make permanent stocks in the herd.

(Festetics 1822)

Festetics' organic and genetic laws, which were focused on *de facto* findings of the biological phenomena of heredity, were accepted by his peers, and they began to use the term heredity (*Vererbung*) in their writings, whereas the discipline was referred to as the history of heredity (*Vererbungsgeschichte*) (Nestler 1837). Another term, *örökldés* (German *Vererbung*), began to spread in Hungarian, relating to the biological reproduction and temperament of traits from one generation to the next (Haubner 1857). His theorems can be traced back to Festetics, and Haubner's explanation explicitly reveals the legal approach used in hereditics (*örökléstan*).

In the nineteenth century, among members of the sheep breeding community, the history of heredity, or *Vererbungsgeschichte*, circulated primarily in the Habsburg Empire's territories. It is also used interchangeably in the German-speaking world with the idea of creation history (*Entwicklungsgeschichte*) (Orel 1977; cf. Shan 2016). There is no clear distinction between development and heredity at this stage of history, just a mixing of theories and explanations; but, in this respect, the two terms correspond to two sides of the same coin. Scholars in the nineteenth century saw heredity as only one stage in the never-ending phase of growth and development, and it never occurred to them to isolate the process of transmission of traits. This concept appears in Mendel's popular dissertation, written in 1866, where he studies the phenomenon of *Entwicklung* (development).

Despite the fact that the word "genetic" exists only once in Festetics's work as an adjective, it does not propagate in the context of a noun, as "genetics", among animal and plant breeders. Festetics, on the other hand, did not use this form neither. We can only guess on this cause without further historical data. Festetics's genetic rules aren't genetics in a modern twentieth-century sense; rather, they're part of genetic prehistory, since variations among farm animals, trees, and people, according to Festetics, are all the product of his empirical rules what he deduced from his observations.

These observations are somewhat parallel with the history of chemistry. Festetics and other sheep breeders, though about plants and animals, through a mechanical lens. Living creature were regarded as automatons and members of the society were following André's ideology who was looking for Newtonian theoretical explanations about the mystery of heredity. He also stressed the application of strict scientific terminology applied in mineralogy. There is a striking similarity how Central European breeders develop further the Bakewellian methods to separate, combine hereditary substances passed on from one generation to another through blood. Later, this is getting experimental endorsement, which in turn results in further experimentation and mathematical evaluation of segregation of general traits in sheep. It is easy to believe that entire scientific fields develop out of vacuum, but this is hardly true. It is also misleading to think that the mentioned parallel chemistry solely developed out of alchemy. Ihde (1956) demonstrated very well that it was rather cross-cutting through different domains: medicine, alchemy, and metallurgy. This was later elaborated by Westfall (1971) showing that the ideas streaming from iatochemistry lead to the formation of mechanical chemistry. Though, many technical terms such as "spirits" referring to certain organic products of distillation are stemming from the Paracelsian words.

Catherine Wilson (1995) has showed that the discovery of the microworld, as well as the obvious function of living animalcula in generation, contagion, and illness, provided scientist of the sixteenth and seventeenth centuries with the challenge of reconciling life's ubiquity with human-centered philosophical structures. It was also a source of consternation for theorists interested in essences, qualities, and the boundaries of human intelligence, whose views are reflected in contemporary discussions regarding realism and instrument-mediated knowledge. This is also true for the members of the Agricultural Society, who constructed two specifically designed microscopes to study the microstructure of sheep wool and assign them in to ten different grades. Recording assessing these traits through several generations can be regarded as a first attempt to investigate the complex problem of heredity following in the footsteps of Maupertuis. Though they failed to understand or explain this complexity in detail.

Why Was Festetics Forgotten?

Research done in the framework of the Brno SBS, which reached
its climax with Festetics's article of 1819, initially focused on sheep
breeding but later extended to the nature of heredity, raising
questions for the next century to resolve and providing perfect
preliminaries to Mendel's work and successive genetic research.
These sheep breeders were ahead of their times even in terms of
research methods, following a principle similar to that of contemporary research networks in the study of a question equally
significant for science and practice. Their "networked" activities
are proven by the extensive correspondence of SBS members that
has been preserved at various national archives, which offers us
a glimpse of the scientific spirit thriving across Central Europe
represented then by the Habsburg Monarchy.

Therefore, we can conclude that their "inter- and multi-disciplinary"
findings synthesizing work in several fields of science could not
and did not fall onto fertile soil in the early nineteenth century,
as the public was not ready to understand Mendel's achievements
46 years later either. To use a Mendelian analogy, the cross-
pollination between different disciplines is what we know now
as "innovative" thinking. For this very reason, multidisciplinary
approaches to research are increasingly popular and supported
(Herrera et al. 2010). However, if their research was so innovative
and ahead of their times, we can reasonably raise the question as
to why we have not heard about them and why Festetics's findings
have been forgotten.

This question can be fruitfully examined from the perspective
offered by Michael Polanyi (1891–1976) and Thomas Kuhn (1922–1996).
It was first Polanyi (1958) then, following in his footsteps, Kuhn
(1962), who made the term "paradigm" a part of scientific terminology. According to their interpretation, a great advancement
occurs in science when an established theory or "paradigm" can
no longer incorporate "facts" accumulated in a specific field of
science. These facts normally produce incremental progress but
there can always be findings which are hard to fit into the prevailing scientific explanation. For a while, scientists struggle to
fit these facts into the existing theoretical framework, until the
data produce an entirely new theory (paradigm). If this occurs,

everything is rearranged, things will have new interpretations, which indicates the beginning of the next growth phase of science. The old model becomes a subset, and a scientific revolution takes place. Obviously, this happened when, having stated that the similarity of species can be attributed to their descent from a common ancestor, the principles of evolution had been accepted. Accompanied by opposite views, this naturally provided an explanation for many things and resulted in a paradigm shift. Suffice it to cite a famous aphorism from Dobzhansky (1937): "*Nothing in biology makes sense except in the light of evolution.*"

Today it is an interesting phenomenon that people who question the relevance of even the most fundamental scientific achievements can easily unite via social networks. For example, they claim that vaccination has no use or importance in health care or that the Earth is flat. I am convinced that we scientists must take some of the responsibility for this, when we cannot clearly explain the new findings of our fields to the general public. We use increasingly complicated and complex languages and terminologies, which are often unintelligible and in turn make some members of our society feel excluded. It is always more agreeable to seek the company of people with similar mindsets than fear the unknown.

Presumably, the lack of understanding and resentment to scientific findings that often seemed complicated and meaningless had a role in preventing the works of both Festetics and Mendel, as opposed to the theory of evolution, from inducing a scientific revolution. The spirit of the age, defined by natural philosophy, did not favor the experimental science these two men pursued, and the contemporary scientific community was not mature enough to adopt the (non-physiological) genetic thought, thus both Festetics and Mendel had sunk into oblivion for a long time.

In terms of the contemporary scientific community, this is a good moment to turn back to the Metternich-era and its secret police, which aggressively and effectively eroded the most important hubs of the emerging scientific networks in Central Europe. We cannot claim that the then circle of experts had no access to Festetics's paper, since the journal *ONV* was highly appreciated and available for everyone interested in this topic. Festetics had the power to convince and argue clearly, which had also been noted by his contemporaries, such as Johann Karl Nestler

(1783–1842) and Cyrill Franz Napp (1792–1867), who later taught Mendel.

In 1819, when Festetics's work appeared in print, and in 1865, when Mendel made his findings public, there was no critical mass of facts in the Polanyian/Kuhnian sense to trigger a paradigm shift in a specific field of science. The political environment of the age hindered rather than facilitated a scientific breakthrough. Again, the extent to which this posed an obstacle should be the subject of further discussion and research. As I have attempted to demonstrate, C. C. André was forced to leave Brno, which we can reasonably attribute to political factors. He certainly did not intend to depart from this city and give up his scientific career. All this had implications for the appraisal of Imre Festetics's work.

The Rediscovery of Imre Festetics and the Central European Sheep Breeders

The fact that Festetics's 1819 paper "On Inbreeding" was relegated to oblivion for the 170 years following its publication may be regarded as one of the greatest mysteries in the history of biology. It was only from 1989 onward that the importance of Festetics's work began to be acknowledged in both international and Hungarian terms. His current appreciation in Hungary owes much to studies by Attila Szabó T. (Szabó T. and Pozsik 1989, 1990; Szabó T. 2009, 2016, 2017), while in the international context the former director of Brno's Mendelianum (Mendel Museum), Vítězslav Orel (1926–2015) and the researcher of University of Manchester, Roger J. Wood conducted in-depth inquiries into this topic (Orel 1973, 1974, 1975, 1977, 1978, 1983, 1997; Orel and Wood 1981, 1998, 2000; Orel and Fantini 1983; Orel and Verbik 1984; Wood and Orel 2005; Orel and Peaslee 2015), but the name of Imre Festetics emerges first in this story around 1989 (Orel 1989).

However, this could happen only after the period between 1940 and 1970 due primarily to the dominance of pseudoscientific views in genetics, which also affected judgments on Mendel's work—again, further investigation is needed to uncover the reasons. This is a long story with far reaching consequences, affecting even Orel himself. The word "science" derives from Latin *scientia* (knowledge), and it often refers to the great enterprise that "gathers knowledge about the world and condenses the knowledge into testable laws and principles" (Wilson 1999, p. 58) through observing, identifying, and, based on experimental investigation, interpreting phenomena. This is the hallmark of empirical science, which is guided by exact sensory observations and experimental evidence (Sober 2008). At this point, it is worth quoting the words of American astronomer and astrobiologist Carl Sagan (1996):

> Science is more than a body of knowledge. It is a way of thinking; a way of skeptically interrogating the universe with a fine understanding of human fallibility. If we are not able to ask skeptical questions, to interrogate those who tell us that something is true, to be skeptical of those in authority, then, we are up for grabs for the next charlatan (political or religious) who comes rambling along.

This clearly shows that, very often, experimentation and social milieu are not compatible with one another. Eventually, scientific investigation cannot happen without the blessings and acknowledgment of society and without the necessary funding. Science is not a culturally isolated activity. Research is organized by norms and values which serve as a basis for appreciating scientists (also as individual personalities)—a conclusion also supported by decades of research findings in the fields of history, philosophy, anthropology, and sociology (Castoriadis 1991; Weingart 1999; Penders 2017). Even seemingly innocent primary school education may focus attention on scientists and researchers who are "worth" remembering (Wolfe 2018). This seems to be particularly true in the case of genetics, a science whose rough history is full of forgotten researchers and their often-unbelievable life paths.

As Attila Szabó T. highlights in his theory, the essence of genetics was born in Central Europe, where the cities of Brno and Kőszeg were prominent locations (Figure 4.2). In the course of history these towns came under political influence that did not favor this discipline, which had serious implications on access to and the study of documents, letters, and notes that survived in this region. For instance, general knowledge of the motivation for Mendel's work had remained a puzzle and mystery for

FIGURE 4.2
Photo of the courtyard of the Festetics palace in Kőszeg, Hungary. (Source: Courtesy of the Institute of Advanced Studies Kőszeg. Photo by Bence Gaál.)

decades. Mendel's figure attracted more questions than historians of science could answer. The same applies to Imre Festetics. Consequently, research in the history of science has to work off centuries of backlog in this area and find the answers to questions in the forthcoming decades, which will demand painstaking and thorough work.

I should explain the above digression through the 1940s example of Trofim Denisovich Lysenko (1898–1976), who managed to gain support from Stalin, then leader of the Soviet Union, for his views of inheritance, in which Soviet propaganda saw a great opportunity. Engaged in plant production and breeding, Lysenko could reach an outstanding result by the preparatory treatment (vernalization) of wheat seeds, which allowed winter wheat to be planted, instead of spring wheat, in places where it seemed to be unimaginable earlier. This achievement was of utmost importance for Soviet agriculture, which was struggling with serious problems and was expected to rise by relying on Lysenko's findings. Therefore, after World War II in the countries under Soviet influence the discipline of genetics had been replaced by Lysenkoism, a pseudoscience overburdened with neo-Lamarckian dogmas. Ignoring all reasonable arguments and acting as a dilettante, Lysenko rejected the existence of Mendelian inheritance, chromosomes and genes as clerical (Mendel) and capitalistic (Morgan) conspiration in science. Instead, he embraced the works of Ivan Vladimirovich Michurin (1855–1935) and highlighted the role of capacities of organisms to assimilate external (environmental) conditions rather than use internal (genetical) hereditary traits. Lysenko believed in the inheritance of acquired characteristics and thought that specific traits were also influenced by how living beings responded to environmental conditions, which had nothing to do with genes– that was considered only a product of Western propaganda. In essence it was a return to the Festetics vs. Ehrenfels debate (1819), but accepting that heredity is guided by physiological laws.

This had a tragic result, claiming millions of lives due to the famine caused by poor crops, which Soviet propaganda attempted to deny. Ultimately, the pseudoscience of Lysenkoism had been paralleled with cruelty in Soviet states, and its opponents ended up in Gulag camps. Maybe the most famous scientist who fell victim to this calamity was the renowned botanist and plant geneticist Nikolai Ivanovich Vavilov (1887–1943), the founder of the science of genetic resources, who formed a theory on the centers of origin of cultivated plants and envisioned a global seed bank, his huge collection serving as the basis of a future gene bank. Vavilov rejected the dogmas of Lysenkoism, for which he was arrested and sentenced to death in July 1941, later commuted to 20 years of imprisonment. Vavilov was conducting field work in a border region between Russia and Eastern Europe when he was shoved into a car, appearing out of the blue, by KGB agents, imprisoned, and later sentenced for conspiracy to death.

Ironically, Vavilov's prison cell was not far from his family and colleagues, who knew nothing about his mysterious disappearance. In his prison cell Vavilov had been trying to stay alive by eating frozen cauliflower and molded flour for 18 months, but finally died of starvation on January 26, 1943. It showed the commitment of his associates, who all believed in Vavilov's achievements, that they strove to protect the plant samples consisting 250,000 items held at the seed bank of their institute in Leningrad (Saint Petersburg) even during the siege of the city. At that time this repository was considered the largest and most invaluable collection of its kind. The scientists put the seeds in boxes and stored them in the cellar, then alternated in guarding its door. By the end of the war, nine of the "guards" had died of hunger, but the collection of seeds had remained intact. These scholars chose to die of starvation rather than to survive by eating the seeds and bulbs. Lysenko's deeds resulted in a severe lapse in science and agriculture, which has been discussed in several revelatory international studies (Caspari and Marshak 1965; Joravsky 1970; Hagemann 2002; Hossfeld and Olsson 2002; Roll-Hansen 2005; Gordin 2012). Here it should be noted that, while the impact of Lysenko's ideas on the history of genetics has been minutely uncovered in another prominent city of genetics, Brno (Matalová and Sekerák 2004), the same work has remained incomplete in Hungary (Fári 2017).

In today's modern biology, epigenetics (not identical to the epigenesis of the seventeenth and eighteenth centuries) focuses on changes in the gene expression of progeny caused by the impacts of environmental factors on parents. This raises the question whether epigenetics can rehabilitate Lysenko's activities. The answer is a definite "no." Although Lysenko studied vernalization,[1] a real scientific phenomenon, he had gravely mistaken ideas of this method. He believed that vernalization (or its Russian equivalent, yarovization), the vernalized state, was heritable across generations. His false assumption caused millions of deaths, which cannot be justified by the existence of the science of modern epigenetics. Lysenko's theories were so horrible that they deserve to be encapsulated in the term "Lysenkoism" forever. This terrifying ideology also prevented, among others, documents concerning Mendel from being processed from a perspective of science history, since it ran counter to the pseudoscientific dogmas prevailing in Czechoslovakia in the mid-twentieth century. As to the rehabilitation of Mendel, the exemplary work of Vítězslav Orel (1926–2015) is outstanding. The greatest features in Orel's oeuvre were his efforts to have Mendel's work acknowledged in communist countries as well as globally and to shape the narrative of science history, whose difficulties were summarized by Paleček (2016) after Orel's death. Like Jaroslav Kříženecký (1896–1964), Orel had a key role in organizing the 1965 Mendel Symposium in Brno, and he contributed significantly to the general rehabilitation of the history of genetics. Kříženecký was imprisoned twice by the Gestapo during World War II. In the post-war period he drafted a plan to establish a scientific institution in Brno whose task would have been to maintain Mendel's heritage and have his achievements remembered globally. However, after the Sovietization of Czechoslovakia, in 1948, Kříženecký was removed from his university position, while Mendelian genetics had been banned and replaced by Michurinist biology. Although he lived and worked in Moravian Brno for most of his life, Mendel was born in the Silesian German-speaking village of Heinzendorf (now Hynčice, Czech Republic), for which he had been vilified during the Soviet period of Czechoslovakia—an inevitable impact of Lysenkoism experienced in all countries of the Communist Bloc.

Under the communist regime, Kříženecký had to face further hardships in persisting with his work. In 1958, he was sentenced to jail after it was revealed by Czechoslovak authorities that he advocated the rehabilitation of Mendelian genetics. In the same period Orel was dismissed from his job as department head at the Poultry Research Institute—probably because of his contacts with Kříženecký. The official reason for this move was that he seemed to be an extreme "individualist." Meanwhile, in the Soviet Union scientists like Pyotr L. Kapitsa (1894–1984) and Andrei D. Sakharov (1921–1989) started to take a stand against Lysenkoism. The failure of agricultural policy, heavily influenced by Lysenko, was one of the reasons for removing Khrushchev from his office (Khrushchev 2007). Loss of freedom had taken its toll on Kříženecký's health, yet he did not give up the idea of founding, together with Orel, a Mendel memorial institution in Brno. Inspired by a visit of American geneticist Curt Stern (1902–1981) in 1963, they began to prepare an International Mendel Symposium to celebrate the 100th anniversary of Mendel's discovery. Thus, in 1965, initiated in Brno, the city where he lived and worked, Mendel's discoveries were rehabilitated throughout the Communist Bloc—a symbolic closure to the era of Lysenkoism (Orel 1965, 2005).

All of the above may give rise to the question whether, in the same way as Mendel's, the rediscovery of Festetics's work was hindered by the political environment prevailing in the mid-twentieth century.

Note

1. Some plants, such as members of the *Brassicaceae* family like thale cress, *Arabidopsis thaliana* (L.) Heynh., which was introduced by the famous Hungarian geneticist George P. Rédei as a model organism of plant genetics, require low but positive temperatures for their initial development. The effect of this treatment varies across species, but it is essential for some plants in order to stimulate the development of reproductive organs or the seed stalk. In the case of wheat, seeds must be exposed to a longer period of cold, which ensures their later flowering and seed production.

Summary

The advancement and flourishing of contemporary sciences rest on collaborative research networks, which can be frequently spotted at surprising junctures within the history of science. In light of this, Mendel cannot be considered a lone friar because, by the time he focused his attention on the issues of heredity during the experiments with self-fertilizing peas, many Central European sheep breeders had already been engaged in the scientific inquiry of heredity for 50 years. Moreover, some of the principles of inheritance described by Mendel had already been articulated a generation earlier in Brno by a count from Kőszeg, Imre Festetics. Festetics must have been confused by the complex transmission of the traits he studied. Now we know that multiple genes interact in controlling the transmission of traits, which define wool quality in the process of polygenic inheritance.

Historians tend to adopt one of two distinct views on the course of historical events. The ideas of these two schools differ in terms of whether they attribute the shaping of historical events to cultural/historical contexts or to outstanding historical figures. To put it in a slightly different way: do these events occur due to the workings of culture or eminent personalities? Those who advocate the theory of outstanding historical figures claim that Mendel's milieu had no role in his achievements, so he must have been a lone genius. In a sense he was by all means an exceptional intellect, since his talent, training, and experimental methods are undeniable, while the claim of his being "lone" is certainly dubious. In my opinion, the truth dwells in the mixture of the two views: while environmental conditions provide fertile soil for the emergence of new theories and movements, those can only succeed if outstanding historical figures, acting as catalysts, trigger change which is in turn shaped by a specific context.

Both Imre Festetics and Gregor Mendel were such outstanding historical figures, acting as catalysts for events in the making; although they did this in different times, they had the same subject: they were both founders of the study of heredity, which later become known as genetics in the twentieth century. Festetics's ideas were probably inspired by German *Naturphilosophie* that also served as a basis for the Sheep Breeders' Society, which evidently had a great impact on Mendel. The breeders of Brno and Kőszeg who conducted experiments on heredity in sheep, among them Festetics, were precursors to Mendel. Irrespective of whether Mendel knew Festetics's works or not—it is highly probable that he did, but it will remain uncertain perhaps forever—both of them were produced by the same intellectual milieu. This was confirmed by James Watson who wrote about his experience during a visit to the Mendelianum in 1968, unlocking the mystery as to how Mendel could make a discovery so great in a distant monastery, isolated from all significant scientific centers: Mendel worked in a peculiarly bright environment created by the city of Brno as well as its scientific institutions and societies (Sekerák 2010).

To date, most students are taught that the science of genetics began with Mendel, thus they would be surprised to realize that many of the principles of heredity had been articulated before Mendel was born—in Brno, the very location where Mendel worked, albeit as a result of a study on sheep rather than peas. If we had to name here one person only, it should certainly be Imre Festetics of Kőszeg. The life, activities, and legacy of this mysterious count are still surrounded by many unanswered questions and need further research.

It has been discussed in detail in this book that animal breeders were always interested in the issues of heredity, and they were aware of the important role of parents or "blood" in inheritance. However, until the mid-eighteenth century it was a common belief that climate, soil, etc. had by far the greatest effects on developing the traits of animal's characteristic of a specific region over successive generations. The Scottish farmer James Anderson had different views, since he stated that accidentally produced variations are inherited together with the general characteristics of an animal (Anderson 1796, p. 23). Anderson also indicated that even a breed with stable characteristics can be improved by selection. Although it was known that various breeds could be improved through crossing with imported animals that had the desired traits, the gradual degeneration in successive generations caused by newly introduced traits was interpreted as a proof of the dominant influence of local conditions or the influence of "pasture." This prevented further breeding experiments, and the advantageous traits that derived from crossing were considered temporary.

This view began to change in the late eighteenth century, as some breeders achieved significant successes in producing commercial breeds for their meat or wool. The traits of breeds they produced through the application of groundbreaking methods (inbreeding) appeared in successive generations without degeneration. Of these breeders, English sheep breeder Robert Bakewell (1725–1795) was the most prominent, whose famous barrel-shaped "New Leicester" sheep yielded meat of the highest quality. Bakewell's success was based on an excellent methodological approach: he pursued inbreeding in a closed stock (crosses between close relatives), which helped decide that "breed" has a more important role than "pasture" in shaping the animal's body form. It seemed that, with the proper technique, there was a chance to fix by "blood"-specific traits of animals.

Bakewell earned huge fame, in both national and international terms, by effectively exploiting the power of "heredity," although it had not been articulated yet in a form as we would expect with today's science in mind. However, Bakewell was very secretive about his methods and, although he was highly informed and well-read, he had no scientific ambitions and certainly did not aim to study abstract natural laws. From Bakewell's starting point, to explore this domain he would have needed an organization gathering agricultural and natural scientific experts, thinkers with multiple fields of interest—like the one that worked in Brno. This had little chance in England, since Sir Joseph Banks treated Bakewell with resentment and saw his methods as suspicious. To reach a turning point, Ferdinand Geisslern was needed, who naturalized Bakewell's methods in Moravia, ultimately bringing them to the focus of scientific inquiry.

Meanwhile, the Moravian capital Brno emerged as a flourishing center of wool industry in the eighteenth century, and rose to fame as the "Austrian Manchester." It was in Brno that local breeders founded their organization, the Sheep Breeders' Society (SBS), with its focus on the practical problems of the wool industry. Their specialized journal, *Oekonomische Neuigkeiten und Verhandlungen (ONV)*, published news and articles edited by SBS Secretary Christian Carl André, a prominent representative of natural and agricultural sciences in Moravia. Annual SBS meetings attracted participants from neighboring Hungarian, Bohemian, and Silesian regions as well as Moravia. Members of the association possessed extensive scientific libraries with a selection of the best journals, scientific and cultural works. For example, the collection held by philanthropist and SBS president, Count Hugo Salm, consisted of 59,000 volumes, to which SBS members, including C. C. André and his family, had full access. No wonder André's son Rudolph also later wrote a handbook on sheep breeding.

Another prominent SBS member, Imre Festetics, could also rely on an enormous library housed in his family's estate and castle in Keszthely at Lake Balaton. This library was so huge that a separate wing had to be built. It has survived adventurously and is currently the largest intact aristocratic library in Europe with 80,000 books. This collection contained the works of Young, Culley, Sinclair, and Marshall as well as the county surveys prepared by the Board of Agriculture in London—the same publications which greatly influenced Bakewell. The Brno SBS certainly gathered people who represented peculiarly progressive thinking and many of them were interested in the improvement of the wool industry. Their annual discussions constituted a genuine scientific melting pot for heated scientific debates and intriguing discoveries.

Between 1816 and 1819, SBS members held a series of debates on the inheritance of wool traits (e.g., fiber thickness and crimp, etc.), seeking patterns that would help them unite useful traits in progeny through crossing. The role of inbreeding proved to be the most controversial issue. Austrian breeder Baron J. M. Ehrenfels maintained that, in Spanish Merino breeds, "heredity" was controlled by forces which C. C. André described as "physiological laws." He argued that if Merinos were bred outside Spain then a degradation of wool quality would be experienced, which he ascribed to climatic conditions. Ehrenfels also held that inbreeding disrupted the "principal plasma" of the animal's organization, which directly decreases wool quality.

As opposed to Ehrenfels, Imre Festetics believed that heredity was controlled by strictly internal factors, and inbreeding could be used to concentrate these factors in order to make the transmission of traits more predictable. His assumptions were based on his own practical experiences and observations on Merino breeds.

Among European breeders committed to producing fine wool, quickly rising numbers were interested in Bakewellian methods, which are often referred to with the term "breeding in-and-in." Festetics began his own sheep breeding practices in 1803 and, after more than a decade of experimentation with rigorous inbreeding techniques, he reached a point where he could not buy animals better than his own. His findings—presented at an SBS meeting—also caught the attention of others.

André attempted to resolve the controversy emerging between Ehrenfels and Festetics. He acknowledged the values of inbreeding proposed by Festetics, although he had doubts about the opportunities it could provide, so he asked Festetics to give a written statement, a summary of his views to be published, as did Ehrenfels. Festetics accepted the invitation in the hope that his 15 years of experience as a breeder would be sufficient to support his claims. As a result, his paper *"On Inbreeding"* was published in 1819 on the pages of *ONV*.

Festetics articulated several rules of inheritance, being the first to refer to them as "genetic laws of nature," that is, he was the first to use the concept of "genetic" 86 years before William Bateson used the noun genetics in his personal letter to Adam Sedgwick. Festetics coined the new term in order to make a clear distinction between the rules applicable to heredity or "genetic laws" and "physiological laws." By describing these "laws" he was the first to recognize, empirically, the segregation of traits in the second generation of hybrids. Irrespective of external factors, he also linked heredity to health, emphasizing the role of inbreeding (combined with strong selection) in fixing inherited traits and maintaining or improving the new breeds.

To illustrate his conception, Festetics mentioned various sheep and horse breeds as well as the populations of isolated Hungarian villages as examples. His observations also pointed to important links between variation, adaptation, and development. Moreover, he noted the consequences and role of selection in heredity, believing that variation and genetic laws interacted in natural processes that controlled the populations of various animals, including humans.

We cannot avoid raising the question: Did Mendel (who was 25 years old, when Festetics died) know about Festetics's work? There is no direct evidence to support this assumption, but it is certain that Festetics's paper was available in Brno, and Mendel was meticulous also in his readings. Mendel's law of segregation is essentially a mathematical proof of law "b" defined by Festetics. According to Mendel's law of separation or segregation, the two hereditary factors responsible for a specific trait—that is, the two members of a pair of alleles for a gene—separate (segregate) in the process of producing reproductive cells (meiosis). He proved this when he observed that the traits of grandparents can appear in the second pea generation. Was it accidental, or did Mendel intend to test the results of a previous empirical observation with mathematical methods?

Although the two men were a generation apart, some of the answers to the questions raised by Mendel had been already there, in the library which he used regularly, on a daily basis. For a while, both Festetics and Mendel were members of the Brno naturalists' association (from 1861, Natural History Society), although their memberships overlapped for only a short time, since Festetics died in 1847, two years after Mendel had been elected a member. But Festetics enthusiastically attended the meetings of the association up until his death. In 1865, Mendel read his own paper at the annual meeting of the same organization, and it was published a year later in the Society's proceedings.

Although it is probable that we will never know for sure whether Festetics directly influenced Mendel, both men emerged from the same community, and Mendel may have been aware of Festetics' ideas propagated by his teachers, Nestler and Nap. The *modus operandi* of this community closely mirrored that of modern collaborative research networks, with powerful and well-funded individuals able to bring together scientists from different disciplines to answer specific questions with profound theoretical and commercial implications.

In 1821, C. C. André, the leading figure in shaping the SBS intellectual environment, was forced to move to Stuttgart, and Johann Karl Nestler took over his role as a chief organizer in Brno. Nestler conducted extensive animal and plant heredity experiments and was Head of the Department of Natural History and Agriculture at the University of Olomouc, where Mendel later studied. In 1836, more than a decade after the debate on Festetics's "genetic laws of nature," an SBS meeting was held to discuss "the inheritance capacity of noble stock animals."

One of the speakers at this meeting was Cyrill Franz Napp (1792–1867), who recognized a role for the "inner organization" of animals in determining their "outer forms," and later went on to ask a critical question: "What should we have been dealing with is not the theory and process of breeding. But the question should be: *what is inherited and how?*" This question pointed to the road to be taken in research for Mendel, who had been working on this subject for decades. Relying on the question raised by Napp, Nestler asked SBS members to focus on this issue in their experiments.

Many researchers have attempted to reveal the motivation for Mendel's experimental design. Was he interested in the questions of heredity or simply aimed to create hybrids? Some authors question the influence of Moravian breeders on Mendel but, as it has also been demonstrated in this book, it is clear that many intellectuals in Brno were actively debating theoretical questions concerning heredity and conducted related experiments. For Mendel, nothing in the whole wide world could serve as a better place for conducting his research than the Augustinian Abbey of St. Thomas in Brno. His mentor, Abbot Napp actively promoted the teaching of agriculture and gave related lectures that Mendel attended from 1844 onward.

Mendel's teachers had a strong influence on him, with Napp effectively "headhunting" him for the monastery in 1843 so that he could be engaged in the selected research topic. Napp was particularly interested in resolving the mystery of heredity, and sent Mendel to the University of Vienna to gain specific expertise in the period 1851–1853. When Mendel returned to Brno, the Napp encouraged him to delve into an inquiry on the nature of heredity. Napp and Nestler were principal figures shaping heredity research in Brno. Both had read Festetics's papers and Nestler cited them regularly, thus it seems highly unlikely that Mendel did not hear about these works from his teachers.

Although teetering on the brink of insight, Festetics's work did not immediately lead to a great breakthrough in the history of genetics. Instead, it sunk into oblivion for more than 170 years until its rediscovery. It should be noted that Festetics did not discover factorial or "Mendelian" genetics before Mendel, but he had certainly laid its foundations by introduced the term "genetic" in a hereditary context as early as 1819.

Unfortunately, with a few exceptions, Festetics is rarely mentioned in contemporary scientific literature on the history of genetics. Nonetheless, through his activities in Kőszeg, he made an enormous contribution to the context of intellectual development in Brno, from which Mendel's interest in heredity arose. While Mendel was undoubtedly very talented, he was by no means a "lone genius"—no more than Festetics was. Both men were embedded in a scientific community—an effective research network—engaged in solving the problems of heredity. The work begun by Festetics had seemingly reached an impasse until Mendel arrived at the city of Brno and, whether by accident or design, chose the right organism for the job—peas with discrete characters and shorter generation times than sheep.

Epilogue

The journey of Imre Festetics from Bucharest to Brno through Kolozsvár and Kőszeg

An epilogue should be short. However, while reading this book, I am reminded of a two-century-old story with a half-millennium-old quote from János Sylvester (1504–1552): *"Collect knowledge... because only intellectual treasure can survive death."* It is primarily to begin this way because the author of the volume has collected lasting intellectual treasures in his book. Secondly, because the first Hungarian data on hereditary diseases were noted by Sylvester's circles in Sárvár in Western Hungary. However, these data remained unpublished in archives of Transylvania. And last but not least, because this volume was also conceived at an intellectual meeting of Transylvania and Western Hungary in 2011. Thanks to the author for not forgetting this.

In what follows, I will try to summarize briefly and strictly, in a personal way, what is most important to me in this small volume. There would be a series of references behind every sentence of mine, but the genre of the epilogue does not allow this. As a researcher, wondering about the connections and values described by Poczai is most important to me. And the fact that Central Europe is perhaps an even more exciting field for a geneticist situated in Helsinki than for those who live in Central Europe. Let us stick to this latter approach—in the spirit of the title of this writing.

From Bucharest to Brno—via Kolozsvár (Cluj-Napoca) and Kőszeg. What connects these cities here? Mendel's role and the importance of Brno in the history of genetics are well known, but decades ago in my own lifetime this role was also accounted for as a mystery. Essentially, these mysteries are somewhat surrounding the town of Kőszeg, Keszthely, and Simaság—as well as the sheep-breeding experiments of the Festetics estates. Brno, and its role in the life and work of Mendel was perceived immediately by Bateson, when he and his wife suddenly traveled to Brno on

December 11, 1904. Whether Bateson knew about Kőszeg and the sheep of Festetics, is also part of the mystery.

But why is Bucharest and Kolozsvár in the title? That is what we can discuss in the epilogue. The Bucharest-Kolozsvár-Kőszeg-Brno axis is an important line in hereditary thinking in Europe. Bucharest is a starting point for both Imre Festetics and also for myself. First, Imre Festetics started here as a soldier. From 1782, he was a hussar officer and took part in the struggles for the liberation of the Romanian principalities from Turkish rule. He was wounded near Bucharest and resigned after eight years of service as an officer, requesting his dismissal from military service. He left his regiment on August 18, 1790, in a camp near Bucharest. This injury in Bucharest probably played a decisive role in making this veteran soldier famous not as a general but as a pioneer of hereditary thinking. This was in consequence of starting to read natural sciences, agriculture, and joining the circles of the Moravian learned society in Brno in seeking further knowledge. Second, during communism, Lysenkoism (details in the relevant chapter of this volume), Mendel's classical texts were published in Hungarian in Romania (Bucharest 1976) Bucharest *"The Century of Genetics."* From the outset, this volume played a major role in the good relationship between the "Mendelianum" in Brno and the Department of Botany at the Agricultural University of Cluj-Napoca—specifically between Vítězslav Orel and me—and two decades later, indirectly, at the beginning of the research on the work of Imre Festetics and his contemporary animal and plant breeders in Central Europe.

This brings us to the role of Kolozsvár in the story. This began in 1957 at the time when Lysenko's dogma dominated the education of heredity; thus, Bachelor students in their first year of biology at the János Bolyai University were not allowed to learn about chromosomes in cytology classes. In the same city where József Gelei (1885–1952), a pupil of István Apáthy (1863–1922) and Theodor Boveri (1862–1915), wrote monographs in Hungarian and German in the 1910s entitled *"Longitudinal Pairing of Chromosomes and the Hereditary Significance of the Process,"* which led him to unravel the sequence of events on the formation of homologous chromosome pairing. My distinguished Professor Ferenc Nagy gave me the noble task of presenting the topic to a small selected group of students in the form of a seminar, since he could no longer do so from the department.

This seminar defined my relationship to genetics forever. Five years later, when listening to the criminal lectures on Lysenkoism in Romanian, I vowed to write a book on the history of genetics as soon as possible. That time will come—I said. My best guess is that Orel had the same motivation when he was removed from the department of the poultry research institute in former Czechoslovakia, and was sent to work on a poultry farm for promoting genetics. He was officially termed as an "individualist" by the authorities. The structure and the role of DNA had been known already for a decade by 1962. But that is another story.

However, this provides some parallels for the history of early heredity research at the time of the 1848 wars of independence then spreading throughout Europe. If we are looking for parallels between the 33 years of the Metternich era (1815–1848) and the 22 years of the Ceaușescu era (1967–1989), perhaps the most striking feature of both is the role of the secret police, the impact of its methods on society and science. I myself place the beginning of the Ceaușescu era a bit earlier to 1957 and geographically to my town Kolozsvár (Cluj-Napoca), since this was the time and place where I met the later dictator in person. The Romanian communist party's leadership entrusted Ceaușescu, as the general secretary of the Romanian Communist Youth Association, with the administration of Hungarian affairs after the revolution in 1956, and especially with the gradual liquidation of the university for Transylvanian Hungarians in Kolozsvár (Cluj-Napoca). His actions led to three suicides in half a year in the immediate vicinity of my family. Two of them took place in the family of Zoltán Csendes, the rector of the University, and another by László Szabédi of my father's department. It seems personal, but it's not: my father was saved from committing something similar by the fact that he was sent to an asylum with a complete nervous breakdown. In this political atmosphere, we were (would have been) taught genetics then.

I endured in this "atmosphere" for another 30 years as a researcher. It was not until 1987 that I left Kolozsvár, essentially for the same reasons as Christian Carl André left Brno: censorship on my scientific writings, opening of my private letters, and certain scientific books were banned by the authorities and out of reach for me. I felt to be running out of air. It was suffocating. *Mutatis mutandis* here referring primarily

to the Byzantine-style "showcase policy" of the Romanian dicta-
torship: the internal destruction had to be disguised somewhat.
Sticking to our topic, genetics, thanks to Géza Domokos, the excellent director of the Kriterion Publishing House in Bucharest, I was
able to write about genetics and evolutionary biology and also publishing works and translations about ethnobotany in this period.

By publishing a Hungarian translation of Dobzhansky's *Genetic
Diversity and Human Equality* in 1985, I ran out of air in Romania
forever. In 1986, I still managed to edit and write a preface to
Stefan Imreh's small monograph entitled *"The Chromosome"* under
the professional supervision of our common professor of genetics,
Endre Lazányi. This volume included the work of József Gelei,
and remained the only volume in the planned series on genetics.
The author of the book soon found a new home in Stockholm,
Sweden, and the editor in Szombathely, Hungary. These were
common fates for Transylvanian scientists in the second half of
the 1980s.

Each of these mentioned volumes in some ways are related to
the development of my attention toward Imre Festetics and his
contemporaries in Central Europe. The book entitled *"Century of
Genetics"* including the Hungarian translation of Mendel's pea
experiments and the development of hereditary thinking allowed
me to form a close and friendly relationship with Orel, the director of the Mendelianum. Studying ethnobotany later indirectly
led me to investigate the story of Festetics. I chose Sárvár and
Güssing to study ethnobotany, which were in close proximity
to Szombathely, as well as Kőszeg, Kőszegpaty, Simaság, and
Keszthely, important for the Festetics estate.

In the second semester of the 1987/1988 academic year, I began
my lectures on the history of science and genetics in Szombathely,
Hungary, at the Department of Biology, Dániel Berzsenyi Teacher
Training College. This was the first year that this university-
oriented college offered studies in genetics. I lectured about
theoretical genetics, while my assistant Lajos Pozsik provided
practical seminars. At the end of semester, we took the students to
the Mendelianum. After the collapse of the Iron Curtain in 1989, I
was able to freely meet Orel. We arranged repeated visits including the staff from the video studio of Berezsenyi College led by
László Pehi and we made educational videos about the life of
Mendel. This was the time in June 1989, during our visit in Brno,

when the Hungarian Foreign Minister Gyula Horn and Austrian Foreign Minister Alois Mock removed the last vestiges of the Iron Curtain at their respective borders. The winds of freedom and liberty were beginning to blow.

It was in this spirit that Orel invited me to his office with a mysterious look on his face and asked me what I knew about Imre Festetics. I had heard his name in this context for the first time—I had no answer. Then, Orel showed me the corresponding volumes the *Oekonomische Neuigkeiten und Verhandlungen*, where Imre Festetics published in Brno between 1815 and 1820. It became clear to me within a second that this was a serious matter. We quickly agreed that I would start searching in the Hungarian archives, since I can read them, and publish on this topic in Hungarian, while Orel researched in the German sources. Then we would share what we had found. The work seemed quite impossible in terms of the large amount of material and resources available. We also agreed that Imre Festetics would be featured in the video then being made.

This brings us to the story of the last 30 years, the period between 1991 and 2021, and to the story summarized in this book. If anything is missing from this summary, it is perhaps the fame or "cult" of Imre Festetics in Kőszeg. It is also important to mention the memorial lectures held annually since 2015 in his home village—Simaság. Even though the tasks ahead of us in this research field are still enormous, this book is an important pillar in upholding the pre-Mendelian heredity research. Perhaps this epilogue was a bit long. Still, it might have been worth reading. What is really worth thinking through carefully, and repeatedly, is the lasting intellectual value gathered in this book, which will survive us.

Attila Szabó T., DSc
Professor of Biology

Appendix 1—The Opinion of J. M. Ehrenfels About Inbreeding

1817. *Economic News and Announcements* 11:81–85

43. Sheep Wool

About noble sheep breeding in relation to the distinguished Ehrenfels-race. To the editor. Supported with wool samples, which are displayed by the publisher in Brno.

(Compare with Nr. 1. 1817)

One may ask from me about the conditions and principles that I apply concerning my sheep, since in the last 15 years so much has been rationalized and irrationalized, publicly and privately. Here I state the following with confidence. **For 25 years I have been raising sheep with some preceding knowledge. Without arrogance: if you trust nature and work with her side by side for so long, sometimes she reveals some things, which remain eternal secrets for others. Independent thinking is an ability, which despises blind imitation.** In the first few years, it immediately became clear to me where we need to bring and lead the fading of wool refining in our homeland, while discussing important issues in agriculture, trade, and production, I have now fulfilling this task myself:

1. A breed of sheep must be reared which, in terms of wool, is not only more delicate than the Spanish, but is closer to the well-known exceptional cashmere wool.

2. This wool must be suitable for making both the cloth and all the finer products.

3. Especially in the case of animals, after finishing the wool, we have to get to the point where we can differentiate between Spanish or, for example, between superfine, primary, secondary, and tertiary wool in terms of production, regardless of the quality of the wool, whether it comes from the back, the heart, or the sides: the shades of

the best quality are invisible to the eye and only after sitting in the factory chair they will become clear.

4. Quantitative wool obviously cannot have such a perfect silk fineness, but if it does not at least on average, it should provide the country with good quality that can be shed in farms.

5. The bodies of our sheep should not be pampered with dietary stimulants and substitutes, the general spread of which is undesirable.

In order to achieve all these goals, the purpose of my lively action was to find an animal body capable of fattening, whose body I had enlarged and whose best wool had been extended. How much I have accomplished so far for these many purposes is evident when some of my wool patterns, and later my live rams and ewes, were praised by impartial connoisseurs in Brno, Meidling, and Schönbrunn alike based on commentaries. As early as 1808, as a result of my knowledge, I had a publication on improving sheep breeding (currently in the press of Heren v. Rössle in Vienna) in which I dealt in many ways with the question of what true wool breeding is.

Some have challenged the principles I have set up there; but they could not substantially refute them, as I will soon present in a few sentences below. Others have muffled their views on my principles, poor in their souls, **who do not understand certain arguments of Nature and its theories, or on the contrary, misunderstand them and want to dispute my principles.** Again, others praised my darlings in exaggeration, so they became almost suspicious of me. I have no reason to be the subject of praise and assassination, and **so I fell silent as I quietly continued my breeding business, trying to realize what Nature had taught me and but she also given me hope. What I have to cite as evidence for the achievement of the five superior refinement principles is as follows.**

First. You and true connoisseurs can judge the characteristic fineness of wool from the attached wool fiber samples from (1) ewe and (2) Ram. Spain also has no finer wool, and this degree of fineness is present only in some individuals and far from all individuals. In fact, as long as the refining is compatible with all other secondary purposes, this type of wool is also close to that of cashmere wool. The degree of refinement is compatible with the degree of Swedish wool culture. However, the economic advantage of production is only at this point and no longer compatible with processing. Achieving a higher degree of wool fineness is still possible, but it can only be used for one purpose, namely to quickly regain our declining role in the competition refinement. This wool, which I have shown in the samples, is an excellent for garments and, at the same time, remains extremely useful for work requiring the most excellent light wool. At this point, the wool from the fine forehead and calf became prima wool; so far, the amount of wool has not been damaged: **at this stage of refining, all the purposes of the wool culture can still be combined, the final wool can be modified up or down in one step, and both the fineness or applicability and quantity of the wool can be influenced.** So much for my experience, and let's get to the somewhat delicate solution of the problem right away: where is the place of the economy in this, and where is the real limit of refinement?

Second. It is true that this wool is particularly similar to cashmere and so on.* But that it is still particularly outstanding and can be used immediately for the production of garments as proved by Mr. Maro from Klagenfurt, who for years has been processing not only my standard Brno wool but also some of my competing sheep, making notoriously excellent clothes. Making them of first-class quality that demands the finest dominant wool, this practical demonstration of the applicability and usefulness of a garment is far more eloquent than any argument in which we try to prove that thicker wool is better suited for the manufacture of garments than finer thread.**

---

* In 1809, the Viennese manufacturer Ritter made a white narrow neckline from this wool, which was as soft and flexible, that as the conspirators would have hidden it in their small coat pockets. In the presence of living witnesses, I declared that I have never before saw wool that would have given white work so much natural light, such fine fabric during spinning, and traits that would satisfy all requirements after the work was finished.

** Read the instruction by Mr. von Moro in the issue of October 1816, page 405. Although I am so far wrong in this comparative judgment because Mr. Moro probably saw my sheep farms at Brno personally, praised and mentioned them, but did not see my so-called race sheep farms in Mangelsdorf and therefore did not mention them, so as not to say anything against his convictions and Germanness.

Third. As for the fine quality of the wool, which I have given as the third principle of my breeding: by this, I mean nothing more and nothing less than that there is nothing else in all parts of the body and on the limbs other than primary wool: the secondary and the tertiary wool completely disappear. Fully extended wool fibers reach all the way below the knee, which, in terms of fineness and length of wool, belong to the category of prime wool. The practical results confirm me here as well. In 1813, I sold some 100 center wool of mine to Mr. Moro. Although they were sheared by the shepherds of the Brno government, they still did not suffer a great deal of deficiency, they gave the whole breed either excellent and prime wool, but they were zealously processed as secondary and tertiary, this was admitted to me by Mr. Moro himself. During shearing, only some parts silt up by the urine were excreted. Foot and abdominal wool, back and rumps were uneven.—**Anyone who asks more in delicacy in wool should reach into nature's magical circles and fetch something from her most secret workshop, which is still not concluded!** Mr. Moro has read these pages and knows and respects these lines as a masterful revelation of truth written in German. My Kenter breed sheep are considered evidence; though one may want to investigate to see if one can find secondary or tertiary wool on its forehead or rump. But, are the Spanish divide their wool for example to primary, secondary, or tertiary grades? Spanish may not do that, but not us. As long as we have to draw the lines which, according to Mr. Petri, designate primary, secondary, and tertiary wool on the sheep's body. I will be also touching on this subject below.

Fourth. In pasture sheep farming, including the lambs and yearlings, I shear an average of 2 1/2 pfennigs of wool washed and cleaned from their body; 3 1/2 pfennigs from domestic ewes. From their limbs or from above their hearts: who cares more?—I also have an excellent amount of wool in individual in my flock: Stöhre, who give 8–9 pounds of sucking mothers, who give 4 pounds of wool washed on the grief. In this case alone, on average, I only get over 2 1/2 pounds in good years, and a bit lower for this weight in bad years. I have often seen scarcely 2 pfennigs shear in noble shepherds, which treat the amount of wool as their first quality. Today my profit included 15 pfennigs from rams, 6 pfennigs for mothers resulting from the trade of characteristic wool. With further completing the sheep one can undoubtedly produce more than 2 pounds of wool on average by balanced feeding, lengthening the wool, increasing the size of the body, and by continuing homogeneous pairing. **I have worked on this aim myself, without the prejudice to the will of greatest fineness and I am happy to admit that I have not yet completed this process, combining quality with the greatest amount with utmost precision. Nature has its own directions in this regard: if you allow yourself to be tempted to go too quickly for the amount of wool, it punishes you with loss of delicacy.** Those who strive for delicacy alone lose quantity. **Nature prescribed amount and intensity in this point in clear terms. Fineness is limited also to measures of quantity, Nature wants to keep it this way, and I could not find the point where both traits can be met at the same time and combined more magnificently than in the available wool samples.**

Fifth. My sheep originate from diverse sources from the Orange breed to the Bukovina. Wherever they are herded, they graze as animals kept outdoors; they are not disturbed by food supplements. **Straw and hay, clover is the food I give to my animals, and although we can sustain and feed the sheep with other things, it is still essential not to give up their well-known food; otherwise they may have a tendency to attribute secondary traits, which can be generated by genetic force alone.**[1] Even my barns don not have a bright look, and they only contain the basic tools and things that sheep absolutely need. I want to suggest with all this that the refinement of competitive sheep breeds dependent on homogeneous mating, healthy and sufficient food.

Now, that I have gone far beyond the 5 goals of cultivating my refining. May one wonder where my sheep originate from? Has this question become fashionable nowadays? Private interest attaches quite special value to this issue; surely, because anything that can ever attain the prestige of sheep of direct Spanish descent can gain a sovereign advantage in sheep breeding, regardless of what the truth is and sometimes the appearance requires, blind faith. It is believed that buck of Spanish origin is a universal tool for any refinement. One forgets that the main certificate of origin is the fineness of the wool and the end of the wool, and the pure data of origin is again based only on the communication of others and it is only in good faith that it is decided whether it is worth propagating everything that comes from Spain. **Let me reverse this question: do people ask and listen to Nature, which is the only real document of origin. The nobler, of course, can** only come from the noble; which is not average; and better not worse. Sheep have clearer pedigrees than any other animal— and will remain so. Fine wool could only enter Europe through Spanish sheep. All wool produced in Spain comes from the same source. Even the so-called Neapolitan herd, which also came from Spain.—With the exception of the Spanish breed, we do not have sheep breeds with which wool can be refined. In fact, sheep considered much better by barbarians do not correspond to our wool culture in general. Sheep from barbarism tend to merge with wool circulating in ports; if our wool is mixed with such wool, it would be much coarser than one might expect, and thus we would lose both the quality and applicability of our wool at the same time.

Therefore: whoever owns wool with superfine characteristics it must came from the Spanish race; because nowhere else is this type of wool found than in Spain. This kind of perfection that we expect from wool, both from the front and back, can be obtained only and exclusively from the Spanish sheep culture. **Anyone who can reproduce this perfect delicacy on their own property without climatic degeneration has certainly confirmed the extract of the noblest original sheep originating exclusively from Spain,** because only and exclusively well-used and mated original Spanish sheep have such characteristics. **In contrast, who has a breed with barrel-shaped front and back, which is only secondary and tertiary, there breeding is not yet complete, and the blood of that breed is gradually going down, regardless of whether its pedigree sheep are derived directly from Spain,** because there the excellent sheep may also degenerate into worse during ennoblement. The hallmark of the nature of true and not presumed originality is therefore unmistakably in terms of uniqueness and the defining purpose of origin, which is particularly important for the most sophisticated perfection of wool refinement. We do not need any defining official genealogical records, blind faith, debate, observance, and praise: **Nature itself gives the most spectacular and infallible features of knowledge in breeding.** This is the only true and convincing pedigree of my sheep; I don't need anything else.

Besides, access to all resources was open to everyone. I bought rams from Mannersdorf 25 years ago for 3 and 6 ducats, from the first Spanish sheep consignment under Maria Theresa, which had many fine properties in terms of the complexity of the wool. From my brother-in-law, Count von Schönburg's stock in Sachfen, just as from other sources in the spirit of perfect elegance, I also had the opportunity to bring out the best. Not only it is important whether we have sheep bought from Spain, but also how we used these sheep, which is decisive more than anything else.

Enough: **whoever has the finest and most perfect sheep also has the best and most applicable original.** But, if we accept the Spanish sheep as the sole basis and mother of our wool culture, why should we want to overdo Spain in Germany beyond this point, and not only preserve the Spanish sheep, even refine them? Is a higher training and refinement of the Spanish sheep possible? Advisable? Foolish?

I answer: **this higher unification remains possible from experience and natural historical conclusions; it becomes advisable not foolish, rather necessary because we are experiencing all the climatic drawbacks of Spanish in their offspring.**

Finer wool than the Spanish already exists in this world: the sheep from the Race animals for example have much better wool than the Spanish sheep. **From a natural historical point of view, the Spanish sheep certainly does not descend from any woolen sheep recently found in Spain. Climatic upbringing and favorable generation under Spanish skies formed the sheep from Athenian and African soil. But the genetic force from this, suppressed but not destroyed by a thousand generations, still preserves the original type of formation and, thus it was easily unleashed by the influence of the climate and counteracting pairing, invisibly redeveloping the properties of ancient origin when homogeneous pairs were met.**

Hence, the natural revelation was then initiated: Spanish sheep, originally descended from these animals, in their offspring merged and obtained a more full-bodied perfection than the originals themselves were. "Several sheep farmers in Germany have tested this natural historical conclusion through experience, and I myself can point to sheep that clearly demonstrate this experience." We also see, through our experience that Spanish originals are certainly coarser in our lands. **If we admit degeneration, as experience teaches, does this also mean that an upward development remains possible? As proven by experience?**

(Continuation follows.)

1817. *Economic News and Announcements* **12:89–94**

46. Sheep Breeding

About noble sheep breeding in relation to the distinguished
Ehrenfels-race. To the editor. Supported with wool samples,
which are displayed by the publisher in Brno.

(Continued from no. 11)

**Since the higher and more perfect integration of the Spanish
sheep is also useful and necessary separately, we can derive
further conclusions from it with moral and proven experience.**
This endures no contradiction, sheep breeders, wool buyers, and
supporters in neighboring countries are already gossiping about
the fact that our shepherds are noticeably and visibly dimin-
ishing in delicacy and this is an unfortunate situation: **the new
principles of sheep breeding are to be blamed; but even the
most consolidated mare makes a gradual climatic degeneration
within our sky, especially when we are inconsistent in mating.**
If, therefore, we do not turn and pull back our shepherds and we
continue working according to the known principles of consanguinity: in 3–4 generations, we will gradually lose our desirable
and praised fine sheep, and in proceeding generations they will
be completely lost.

The practice through experience taught me, therefore, how to
achieve the highest and most attainable fineness with a few individuals and to hold onto it for a long period of time. One would
abandon this general expansion beyond excellence from an economic perspective that supplies generally functional wool: **these
are necessary to recuperate from climatic degeneration of the
wool, and to enhance production of the sheep breeder according to the stimulating power of Nature. I cannot give a textbook
long example here, and I can rewrite my experience in a way it
would satisfy both theory and practice at the same time.** Just
because I have already provided explanations about the more
recent methods of higher sheep-breeding; I highlight the most
harmful things and explain myself about them here briefly:

1. The choice of the parent sheep is more influential then
   original descent on wool fineness and perfection of the
   wool.

2. We should raise the amount of wool as the first required product.

3. **I do not recommend unbounded mating in closest blood relationship. Everyone has their opinion and certain tolerance is present in the field of human knowledge and also in religion.**

I can be wrong of course, and I can perhaps advise you are wrong. Herein, my experience ensures me, and on the other hand I always ask Nature, and I have recorded her answers and announcements in her laws several times. Only by choosing the parent sheep we can have the same value (the highest and most perfect fineness of wool). On the other hand, greedy people will still reach out for folds and flaps of the skin instead of wool, and one has partly found the reason why currently wool merchants and drapers complain about the decline and coarsening of some sheep farms. Buying and choosing according to blind authority and just believing in origin and ancestry is no less harmful. Presently, they have come across animal races that were credited as Rambouillet. Private interests and preferences allowed some belongings to blend with originals that are not original at all. In good faith we intend to buy original wool where true refinement goes back at least 10 years with the finest coarse wool, and what we get, wool of a bastard. **If Nature did not have its own invertible and untouched characteristics of original descent, it would be much more difficult to raise sheep!** Increasing the amount of wool to the first refining level necessarily also impairs the fineness of the wool, gradually suppressing it. There are individual woolly animals whose wool fibers on the backsides and chest are quite coarse, mixed with more or less good quality of the wool; they give only first class [from these parts] and only secondary and tertiary wool measured in their whole body, but in different proportions on their own they may also give more wool weighing more than other superior sheep. These sheep are preferred, because they give more wool than others. Is it preferable choosing these sheep for mating? I know that economists prefer whatever brings them more money; I know that the wool merchants are not wool enthusiast, and that they pay ordinary attention to perfection. But if this will always be the case, where will our herds of sheep end up with such methods of reproduction?

Only the amount of wool can be estimated which is compatible with the highest perfected fineness; it is therefore eternally subordinate to delicacy and not the first refinement quality. Even in an economic relationship and according to isolated interests, the quantity of wool cannot be the remaining primary aim of refinement. How are we going to send our wool for longer time? The increased exit tariff in convention coins is not applicable for medium wool, which has already led to the fact that only excellent wool can be exported. **The consumption of wars stops where one could not see the immense magnitude of the need in terms of the higher degrees of delicacy and applicability.** So only superfine wool will be sought for abroad, but the large amount of medium wool is only looking for sales inland, and the price and value of one to the other will quickly become apparent: they are likely to have a 2:1 ratio next year, and if this semi-finished wool is bought by sheep traders they will not find a conventional value below 100 guilders if their latter products are valued at more than 200 florins due to the declining number of first class breeders and then the products will be sold abroad as rare goods.—Where can the economist find a better interest in the greater quantity of improved wool? First, its quantity, which according to its weight will be more valuable in terms of the price and this is not according to its finest quality. This spreads the wool of a sheepskin into the well-known four grades of fineness and, for each value, equates a corresponding monetary price.

On average, this does not affect larger amount of wool. But how does this harm the entire sheep farming? Is the artificial degradation to the amount of wool destroying manufacturing?—In the autumn, I wrote about higher sheep breeding, where first I mentioned the ranking of refinement characteristics of our woolly animals, where the fineness of the wool should be first priority, then length of the wool should be second, while quantity is only the third rank among trait refinement.* **If you are interested in**

---

* On Page No. 34, Rudolph André is quite clear about the instructions for the refinement of the amount of wool in sheep. Among other things, he says: combining the highest amount of wool with the highest fineness has hitherto not been achieved, and by nature it should remain so. At most, if you use a significant amount of wool, you can clearly associate it with second quality. But how can one not go back with this degree of delicacy? C. C. André.

reading the natural historical reasons, you will find it there. Here I mentioned the major point: the striving amount of wool cannot adjust with its delicacy, by choosing inappropriate sheep as parents, is gradually degenerate the fineness even more.

There are ongoing debates about refinement up till this very day; wool is not the first and most excellent refinement property. Unconditional consanguineous mating is the third principle among novel methods, which are detrimental in progressive ennoblement, [and this method] is already devastating our sheep farms. I am brightly convinced of the harmfulness of this principle, and I had my own article printed in the *Wienerzeitung* in 1806 as an enthusiastic friend and promoter of the noble breed of sheep, which I repeated in my first issue on higher sheep breeding in 1808. The principles outlined were criticized, even in these papers, and stated as unacceptable. I allow myself to return to these points on this occasion and to outline more clearly the point of view from which I judge my teaching. In the above-mentioned lecture, I defended the principles of the ancients, namely to avoid reproduction among close blood relationships, which is not allowing fathers or the other way around mothers to reproduce among themselves in the same family line: so I did not mean that we should refresh Spanish sheep with farm-land sheep abandoning the species and genus, and so to speak, mate the Europeans with Moors. Such nonsense cannot be concluded from my thinking; we are also not allowed abandoning the Spanish race, we may refine sheep from below upward, or preserve original strains from above downward, but still less is cultivated from the central characteristics of the Spanish sheep.

It does not follow from this, that we let sons jump on their mothers. This would be contrary to Nature, as I said at the time, the main plasma of animal organization, experience and ennoblement. The experience of B. has proven, and has since been spoken louder than back then in 1806, that even the original Spanish sheep farms have lost some of their delicacy. What is to be blamed?—Consanguineous mating and mating with the resulting individuals, even if they have climatically declined in perfection and delicacy. Under Spanish sky, there in their homeland Nature formed the original external structure of these sheep, here in this place and establishment; where the air and food by all means found to be different to organize a being so tightly similar to the original, thus distributed over different regions sheep start to gradually degenerate because Nature operates more easily to quickly unleash heterogeneous forms and climatically build them up again in German lands. Mating therefore requires less care in Spain and more in German lands. For us this obviously means climatic degeneration, at first invisible, then finally obvious, which in turn affects the delicacy of Spanish sheep and we have to improve them in order to overcome such climatic setbacks, and to preserve the Spanish wool and finally to form a non-Spanish but German sheep. This force shall remain in the animal with, which we cross and must improve our climatic decline in delicacy, instead, allowing the fathers laying with their daughters thereby giving the flow of life formation a misleading direction to purify what is no longer Spanish and already under climatic influence. For natural historical reasons, I have already shown above that we can create and preserve this finest extract from Spanish sheep here; only Nature wants to know how it works according to its firmly established principles, this is what I meant when I announced publicly in 1806: "Mixing and pairing, refreshing the blood through so-called crossing with noble races remains extremely important in our country, and under our circumstances will be forever necessary!" The daughters engaged by their fathers, and mothers jumped over by their sons, are just devastating among each other; [mating] in such close relationship accelerates and generally promotes natural climatic degeneration, and to avoid this we can probably never abandon the Spanish race, but rather within this race we should form stock that are not too closely related, and choose a more harmless reproduction, or choose parent sheep from other lands, which are completely ennobled and do not share such close relationship. This is necessary among entirely perfected Spanish sheep in our climate; but especially needed in case of those originals or their descendants from Spain, which initially did not yet reach such complete fineness.

As I said: as long as we are adhering to the lines of B., and separating the wool to primary, secondary, and tertiary grades according to Mr. Petri—who have also correctly and clearly represented this—then we will have pure, perfectly fine sheep, sufficient races, which have been consolidated within themselves in Germany, and propagated without setbacks, but if we keep on producing animals resulting from completely uniform making, propagated among each other through consanguinity, the lamentations of our wool producers about the raging problems will increase more and more. Yes, even those of Spanish origin can be considered imperfect in this respect that were selected in Spain for their shape and not for the fineness of their wool, and this quality thus remains in their offspring, proving that fine parents have the same children.

According to my old experience, in our climate they go back with a double step without having become mutilated in their blood or propagated in any other way than among themselves. This is why the following principles and basic laws in mating should be introduced in higher sheep breeding:

1. Fine, perfectly complete woolen sheep can only be acquired from a Spanish race.

2. Spanish originals are indifferently propagated among themselves; in our climate they are losing the unity of their wool.

3. These setbacks occur more quickly if individuals are chosen for mating and reproduction in which the focus is less on perfect wool fineness than on the amount of wool and body figure.

4. **In its revenge, the Spanish sheep not only suffer from the climatic hardening of its wool through indifferent or persistent reproduction; it also allows a refinement of its wool through quick selection and pairing of homogeneous animals. Upgrading and downgrading are possible.**

5. **One can raise offspring from Spanish sheep that are like the originals themselves.**

6. Only through this extract of fineness can the original fineness and usability of Spanish wool be preserved for eternity under our sky with perseverance.

One more word about sheep's sweat, or the wool grease, which is found between the skin and wool. **Some attributed this genuinely yellow sweat, which leave several decent traits of the wool dependent on it, to be the resulting characteristic of the**

refinement process of the original Spanish sheep. This is completely wrong. Every sheep's wool has grease, the common as
well as the finest Spanish sheep. This grease is not a race-specific
trait. The stronger or weaker production of this wax of the wool
depends solely on the feeding and behavior. Sheep kept in the
open air sweat less than those kept in stables; leaner sweat less
than that fattened. At most, one can accept superfluous sheep
sweat as a favorable sign of health. My experiences with sheep
sweat also offered me something else. There are a yellow and
a completely white grease. This difference in color was noticed
initially in Spain, since the tribes and their descendants raised
from this flock came from there. The manufacturer Ritter and
other adjusting producers first drew my attention to the essential difference between white and yellow wool grease. For they
maintained: wool with white fat does necessarily need the urine
washing process in the beginning like wool with yellow fat does.[2]
The wool thread would therefore wear out less with the former,
since it suffers less from expansion in its spirit and thereby promote working with it. In addition, this original natural whiteness
takes on delicate colors more easily and more uniformly, and the
more beautiful natural shine is unmistakable in the raw-colored
manufacture.—I therefore decided to cultivate the offspring of
several completely white-greasy sheep from the extremely well-
established sheep farm of my brother-in-law, Count Schönburg
zu Rochsburg,* and in the end, nothing needs more effort and
care than the breeding and maintenance of the sheep with white
sweat, the mating them with yellow grease producing sheep, then
finally blending them for constant procreation.

I would have been less persistent in preserving this white
greasy wool if only nature had not already shown in the second
generation that the highest unity of wool can only be achieved
by this type of white greasy woolly animals combining with the
highest wool for final processing. Immediately the wool becomes
denser and sharper, as the white grease wool turns into yellowish
wool, **which can be seen from the comparison of the adjacent
wool samples No. 1 and No. 2 with No. 3;** therefore, not all white
grease wool is essential. But I am just talking as I am writing this

---

* Read what Mr. Advisor Thaer said in his popular handbook about sheep breeding on page 81, about the Rochsburg sheep farm, and about their highest wool
  consignment. C. C. André.

essay instead of making important notes. My love for this matter must excuse me; I can only harm the reader with a few results that follow from the preceding:

1. I give a faithful view of those principles who originally led me in higher sheep breeding, and I substantiate these principles with natural historical reasons and experience.

2. In my experience, I seek to define the boundaries in wool samples 1 and 2 that a sheep farmer should not go beyond for his own economic advantage, which as a kind of boundary simultaneously offers guidance for all wool types and sheepskins, thus allowing everyone to practice higher level of breeding. My experience so far has had no further results.

3. I have tried to show what are the grounds of true sheep wool ennoblement, and how original sheep worth for reproduction can be recognized. The main certificate of origin, at least for our wool culture, is the perfect fineness of wool: a natural stamp of pure descent.

4. I summarized the results of my experience: that Spanish sheep not only sustain themselves under our climate, but even allow refinement, and I supported this statement with natural history conclusions and empirical observations.

5. **Again, drawing on experience, I offered methods to ensure that the climatic degeneration of Spanish descendants can be stopped by creating a kind of especially trained fineness extract of sheep that is always uniform, which can thus reproduce on their own.**

6. I have tried to prove: that not the amount of wool, but only fineness, are the primary goals of refinement.

7. **I have more precisely defined the teaching about consanguinity and its effect on procreation by clarifying misunderstandings about it,** and finally.

8. I have recorded my own basic experiences I have followed so far about the important doctrine of mating that I myself have mastered in order to perfect my sheep in their delicacy.

9. I also made important concluding notes about the importance of wool grease.

Excuse me in some cases, for a dominant tone of my writing, but that is the language of experience. Here, too, as everywhere, opinions remain divided for a long time. Everyone can afford to talk about this matter after two years of experimentation and they often write about it. It is hard to get along with inexperienced purely theoretical people, and it is just as hard to convince people who lack theory and rely solely on experience. Therefore, let us be tolerant of ourselves and our humble opinions, not every master of mathematics has to be a Newton, not every sheep breeder a Buffon, nor every collector of herbs a Linnaeus. Common sense often sees correctly, without excluding the higher intellect, it can surpass even the highest. I love my homeland and I deliver everything with warm loyalty to make it prosper. In most of my experiments described here, I was seeking to maintain the merits of our country once achieved in sheep breeding. This urge alone, and not the laying of a solid foundation for knowledge, guided me to share my experiences.

Vienna, February 4th, 1817, J. M. Fr. v. Ehrenfels

———

Editor's note: At my request, Mr. Berfasser gave me 40 rams and 40 ewes to be inspected my correspondents or friends. Rams and mothers are one and two years old. I have looked at the wool patterns and I can safely state that it is made up of the finest wool fibers from forehead wool to kneecap. Prices and conditions can be requested from me verbally or in writing. C. C. André.

Appendix 2—Explanations of Emmerich Festetics About Inbreeding

1819. *Economic News and Announcements*
2 (Supplement):9–12

Debates: Sheep Breeding

Explanation of Count Emmerich von Festetics
(Compared Nos. 38, 39, and 55, 1818)

The *Economic News and Announcements* in Numbers 38 and 39 is covering studies in which an anonymous person debates about the Sheep Breeding Association's meetings, and the sheep-breeding community deserves acknowledgments for any overlooked minor errors; in which he, the enduring secretary, allured Hon. Advisor Mr. André for corrections, new observations, which confirmed his zeal for our company, and his profound knowledge; in which he, the permanent secretary, enticed Hon. Advisor Mr. André for corrections, new observations, Advisor Mr. André will publish me as a curious empiric from Hungary, thereby contradicting the investigation of truth there, where I have gathered practical experience, which my published reports have represented; thus, I will be linked to the author of the report, where the worthy Mr. editor makes his 12th note, without those gentlemen, whom this man found worthy of helping, defending their friendship, and with him, bringing much good and at least from that writing generated from their measure which Mr. André wants to be aware of, for which one must also be thankful, for trying to do the least amount of harm.

Of course, in that article, I will designate the groundwork that I have contradicted, which my esteemed friend Baron von Ehrenfels created, and which should genuinely be referred to as solely physiological, and which the publisher has asked me to illuminate and underline with fundamentals also with practical evidence.

The query may be phrased differently. I take it as follows, as stated by Advisor André: **unconditional reproduction by blood in the closest relationships, set according to physiological nature and continuing for a long time, will inevitably lead to organic weakening.** The following questions arise: whether such a replication is often practicable or even necessary; whether it should be advised; and whether the matter of the issue will not be identified.

It should not surprise me that my irrelevant sentences are bound to be strongly questioned. My understanding of natural history comes from occasional reading, and thus I gathered my remarks with great care but I did not find answers to the questions hence I have created my own system.

First. I must confess that I'm not sure if the term "organic weakness" is properly interpreted. Now I add the following concept to it: the subject's inability, despite being otherwise in good health, to function with its own organism according to natural law and to survive for a long time.

Second. Under organic functions I consider all that is for the conservation of its own self, for replication of itself into subjects that are like them, and for achieving the goal and which are able to demand from the subject with intellect according to natural laws.

Third. A strong constitution is appropriate for fitting, long-lasting conservation, which will be partway innate, and will partly increase due to upbringing, indeed which also is able to decline.

Fourth. This consistently robust constitution is unconditionally a requirement, found with it in reproduction evenly, which the elderly originate similar entity. Often fathers produce minor corresponding descendants. It was the constitution, regardless of health, which had weakened.

Fifth. When the father and mother are gifted with the strongest constitution, those qualities which I claim for my objective, then it is a freak of nature, when in descendants the traits are altered, or whenever the ancestors are not equipped with enough of the required traits.[*]

Since I have seen and read so much on and about crossbreeding, I think these fundamental laws are self-evident. As a result,

---

[*] Perhaps it seems something is omitted here in the transcript. C. C. André.

the discoveries that I was able to make in everyday phenomena forced me to determine over any doubt.

Indeed, this natural historical interpretation does not include humans, I reasoned, since the type is propagated during the reproduction in the closest blood relationship. However, I do not want to seem to be impugning Genesis with this, so I would like to add that the individual is taken into account not only in a physical, but also in an intellectual sense; otherwise, even the most intelligent and hard-working person is subject to good constitution on a regular basis! It's even regrettable! That the individual (so-called civilized) has been quite withdrawn from the primitive natural state, where the statements in the 5-points above will have influence, through his upbringing, nourishment, stipulations, and the form of occupation, and measures for sensuality the necessary instinct for reproduction. We still see the persevering embankment of Christendom here, in my homeland as well, in that, fully depopulated by battles, now by shortage of rations of populated areas, distinct indications of tribes, races, or breeds, when one wishes, that one can surely shape an opinion about their tribe from their entity. Hungarians, Armenians, Slovaks, and Gypsies have distinct facial features, structure, and a distinct manner of behaving themselves in certain regions, so that one can quickly and easily form an opinion about their tribe (sometimes called *échappés*[3]). Power, good health, industriousness, perseverance, as well as spirit and bravery are all qualities that such people possess.

There are herds of diverse livestock in my Fatherland, including horses, swine, horned cattle, wool-producing sheep, and poultry. Crossbreeding became commonplace for horses 30 years ago. There are also a few stud farms that preserve an old race based solely on gifts, from which worthless creatures have arisen; these correlate to their size, at best. Our Hungarian horned cattle have remained the same in some condition (without crossbreeding), larger or smaller, depending on what they are fed for improvement. Many types of swine have remained the same due to inbreeding, and I must admit that the accelerated fattening of these animals is an obvious advantage; we have Bakewells among the common people in this regard. Our Zackers (horned sheep) and Zigarros were propagated by inbreeding and will continue to be the same. The brood consists of 24 eggs from nine hens and a rooster, from which 50 proper poultry were raised for housekeeping, and all is so fine, so persevering, so organically perfect that a person could only wish it.

Now I will turn my attention to foreign findings. I am obligated to include excerpts from the perspective of the English, who developed these principles [breeding in-and-in] and who, as such deserving spokespersons, have been preoccupied with great instances of inbreeding since then. I see individual farmyards (from which excellence of the species would inbreed there, extended into the middle of perfecting) in the Ilz Valley, Märtz, Puster, and other valleys of Tyrol, and Switzerland, and the bulls bought from here improve in the second generation after reproducing with first cousins; this is procreation in close relationships, if not unconditional.

On the other hand, I saw wealthy gentlemen with lovely animals producing milk obtained from those famous valleys; they considered all that needed to be done, and they even hired a farmhand from those areas for the job. They abandon the poor animals here to the farm officials because they lack perseverance in one task and quickly switch to another hobby. In scant willow pastures, old and young mixed, the now half-despised animal's endured starvation, thirst, and suffering due to a lack of sufficient and healthy winter and even no summer fodder. It did not fade away quickly, and the younger generation had a picture of suffering in the environment in which it grew up, and reproduction in close relationship, unconditionally, that is, without respect for age, strength, or fitness, was explicit; nevertheless, it is depicted in a different context. I saw rams and ewes, old and young, 40 heads, incidentally to each sex, purchased this year's so-called Merinos, even though they were costly, being watched by the same farmhand. This may be referred to as conditional mating.

After learning about these events, their origins, and the sad or amazing effects, **I devised my own set of laws,** believing that I was obligated to obey them to the letter in my stock-farming. The first step was to figure out what my aim was for each type of cattle. Inbreeding cattle are the way to that end, a judicial crossbreeding in bastards in the race. The aim of horned cattle was milking, as the cattle has a strong proclivity for putting on flesh. For milking, **I have bastard cows that have been inbred until the third generation, and I have created a breed that has given me satisfaction; I was also able to focus on formation and color as a secondary goal. I have obtained 12 head of pure race swine with the desired characteristics; the swine with the shortest snouts for**

reproduction, and the 360 head now born out of this inbreeding exceeds my expectations. The fact that I complied with inbreed ing of the nearest and oldest bond with this sort of animal cannot be denied: moreover, it was not unconditional. Health, stamina, age, and the appearance of values that contribute to my target were all requirements. It was always the intention that during the years of development of the horse, whose constitution may be solid, the attention paid to after-breeding would be similar to that given to domestic cattle. **Adult animals that produced the worse results were often discarded in these conditions, while the perfected ones were retained for further inbreeding.**

Even for me, Patty, one purchased property in the midst of many completely depleted fields and an utter shortage of meadows is a tough lesson because the cool yellowish earth yearned for a warming fertilizer hence sheep manure; thus, sheep breeding. I wanted to develop sheep breeding after the death of my father and a friend, David v. Chernel.* This Hungarian Geisslern instilled in me a desire to take ennoblement to the next level. I was trained by the British, whose worthy editor made good use of his time and moved banned Spanish domestic animals to the Swiss Alps. As a result, I was able to procure 14 banned Spanish two-year-old ewes, 16 Rambouilleter ewes, and 10 rams. I liked the wool from a ram and a Spanish ewe. I'm not sure if I found the right option because I am still undecided on what the right wool is. Enough, I strayed from my preferences, and I have always worked on the best yarn, which seems to me to be the ideal, since 1803.

I did not think it was more certain to continue than when **I would conform with inbreeding in these 10 head, according to the physiological ideas impressed upon me. Just the same selected rams were used in this breed for 16 years.** Inbreeding was not unavoidable. The ram had to be at least 30 months old, solid, and have the highest quality of fur. My people refer to the breed's first father as Crooked (because he limped a bit); his descendants included the Palasor, Araujuez, Mimush, and Little Mimush; this year, the last gives way to his son Paular. This is the ram's family tree, and he is still one of the former sons. On the female line, I must confess, reproduction was done with brothers or nephews. **As a result, that which Advisor Mr. André refers to as a completed race, and to**

---

* An illustrious sheep farmer who is still relatively unknown. C. C. André.

christen my favorite associates the Mimush-Race, is assured. I may point out that, in addition to the special wool, this small herd has developed a distinct shape in which the animal differs significantly from that of the larger herd, a form that is found in all Mimushes; hence, despite the fact that I only have 24 head of two-year-old rams who are true portraits of their forefathers, 16 head should be awarded my next exhibition prize.

I cannot determine to what degree the organism becomes stronger or weaker in a race, and therefore I believe in doing the best I can, whenever opinion goes down, the esteemed society [not only by their inspection committee, but also as carries over in plenum. In order to accomplish this, I will bring Mimush, six years old Race-animals and a two-year-old sibling, to the next meeting in Brno.

Should I also say what kind of wool it was, and why I like it? It was that, in addition to those qualities, which made the fleece fine, on the animal thus shaped stands there, to a degree, by characteristic power of cohesion, fibers mutually supporting each other, solid, almost like asbestos. This consistency always hides the very fine quality to the eye, but it does not contradict it. However, by this, I just meant preserving the wool's thickness and elasticity. This year, when I delivered 6 rams to the inspection commission with a threefold character of the cloth, I was unsure whether I had not compelled mere partiality, around so much more; here I perceived so much worthy of praise, of wool, which thus varied fundamentally from my chosen wool. The decision in favor of No. 9 reassured me, but it did not entirely correspond to my question, which was perhaps not phrased well enough: here, this animal with threefold nature, which always bears wool of value, motivates one to make a decision, which sheep raiser of the Austrian government should take the trouble to call in especially. A more thorough description of the qualities would have gotten us closer to addressing André's query about the right yarn. And we do not yet have in our language the ability that which, by perception, accurate study travels toward an idea, to be clearly printed. Thus, the evidence is irreversible; the association, where various industrious breeders bring their wool together for comparative study, several confused ideas are reported from not calculating use for all of that; they should share their experience and sharpen their meaning in order to make judgment easier (end follows).

1819. *Economic News and Announcements* **22:169–170**

61. Debates: Sheep Breeding

More Explanations of Count Emmerich von
Festetics concerning Inbreeding.

(Supplement No 2—1819 Compared)

Mr. André's remarks, which he considers following my clarification concerning inbreeding good, also in near relatives, in the 4th Supplement of the *Economic News and Announcements*, stem from his restless striving after, which he wants to know the facts elucidated for each significant controversial issue. I happily accept this recommendation and will be present, where I may have been misunderstood or my message was not fully expressed, to add additional stipulations. First, I would recommend that the fourth and fifth lines of the second Supplement, page 20, column 5, read as follows: "**... variations are deviations of themselves; or were not with ancestors ...**" c.c. **In my opinion, these following five paragraphs contain the genetic laws of Nature. These are the points, which must be disputed; otherwise, my scheme would stand. Because I say the same:**

a. **Animals of healthy and robust constitution plant and bequeath their characteristic traits.**

b. **Traits of the predecessors, which are different from those of their descendants appear again in future generations.**

c. **The animals which have possessed the same suitable traits through many generations can have divergent characters. These are variants, freaks of nature, unsuitable for propagation when the aim is the heredity of desired traits.**

d. **Only those animals possessing the desired traits in abundant amount, can be of great value for inbreeding.**[*]

Following these conclusions, I would refer to the argument that **if I wish to keep a certain trait in the animals and make their progeny inherit and continue it, then carefully executed inbreeding is to be suggested, without the risk of organic weakening** being

---

[*] In my opinion, this underlines the main point. C. C. André.

the inevitable result. The analogies cited would not be enough to persuade me to change my mind.

As I previously mentioned, there are significant differences between human and animal reproduction, and intelligence plays a significant role in human reproduction. But I am curious whether inbreeding among humans could be achieved through meticulously studying my conditions? Especially in situations where weaklings congregate and where the heart and mind are forever divided. My strong ram has a distinct spot among his harem, and even then, he can raise lambs that will ruin the race due to inbreeding. There is a cure for these lambs: they can become muttons. **In terms of the horse analogy, I recognize that more subtle traits than those of sheep must be recognized,** but I must state unequivocally that the inbreeding scheme essentially ruined our horse breeding operation. When it comes to sheep, I have to think of health and wool. For the animal, no one has yet articulated what is desired in a horse in a simple and understandable manner. Apart from its utility and service, I assume that. Aspects such as aesthetics play a role. How many different functions have we asked the horse to perform? And horse breeding and the purpose thereof should be equally diversified. **The structure of the horse is a manifold architecture that is difficult to fully understand. This horse machine will have to be mobile,** each would move to its own tempo, in its own way, in this or that changing direction. Could here, the cross-breeding of heterogeneous ones, be the fortunate system? Certainly not! This was confirmed with our mare breeding, which had died off during the last few decades, but which is now being reinstated.

Mr. André agrees that if Nature is left to its own procedures, it stays truer to its inventions, while with humanity and farm breeding, the fundamental force is diminished, and a disposition toward organic weakening emerges. I would not only fail to refute this, but I will also provide a new example for the belief that Man, enslaved by society and business deals, must return to a normal way of life, both in his home and in his care of his animals. I have a naturally wild stud farm with 100 mares and 4–6 stallions on my land. The best collects his mares first, followed by the weaker ones. The preferred mares are steadfast in their devotion to their stallions. The poorest stallions are always left without mares or with those not selected by the other stallions. The best stallion colts have inherited the organic traits, according to the Csikóses,[4] who has been educated in the trade of herding horses for decades. **Only this would be passed on to future generations, resulting in inbreeding.** Our Gulen are in the same boat, as are the Zigarros and Zacklen. **Sadly, the civilized individual always finds himself in the uncomfortable situation of having to choose the least expensive choice. He is unable to discuss samples and options.** Finally, he found courage and purchased a steer as an example. If one must buy, one can choose something cheaper because, in addition to the money invested, it also nourishes more. Better calves would naturally be born, but not as good as those from the bull that grew naturally and was fed better. Traditionally, in villages of my homeland where cattle breeding is conducted naturally, steers kept in stables during the winter are kept together during the spring before being released to graze, so that they can compete between themselves and determine their place in the herd. Several decent steers have been sidelined as a result of this, with the others pursuing him whenever he dared to approach. The standard procedure will then be to market the animal to a butcher.

These traits can also be seen in sheep, where six rams are attached to a flock of 300–400 sheep. One ram was trailed by ten sheep, while another was trailed by just one. **This all supports my belief that in Nature, as well as in our civilized lives, it is not inbreeding that causes debilitation, but rather other factors. However, for those that lack zeal and routine, continuous alertness, scrupulous choice without prejudices, a fantastic education system, and the characteristics of vivacity, fitness, and mother love would be difficult to attain, and I believe and want to reiterate that inbreeding is not recommended.**

Among plants, the female flower is fertilized by the flutter of a moth wing or a breeze; even the placement, location, and consistency of the soil play a role; several varieties are produced this way, and even the best gardener cannot make or prevent it. As a result, I revert to my conviction that if one wants to keep a stock's characteristics, one can never abandon the breed from which the traits originated. The livestock must be bred to have

a strong constitution in order to produce strong generations of this breed. The stud parent would be the one who has the most desirable characteristics. When any of the laws are followed to the letter, there will be no organic weakness as a result.

Naturally, there is one more condition: you yourself will have to work with tireless effort to experience Nature's own prescribed rules.

1820. *Economic News and Announcements* **4(19):25–27**

12. Contributions of the Royal Moravian/Silesian Agronomy Association.

Count Emmerich Festetics' Report as the Representative of the Sheep-breeders Association of the Eisenburg Committee

Highly Esteemed Associates!

I will present my impressions of innovations in the field of wool improvement and the spirit that rules this company in order to serve my duties as the representative of this country. **The desire to increase the consistency of wool is omnipresent.** Furthermore, knowledge that this gentle, agriculturally most productive animal requires better fodder and improved shelter has grown. As a result, the fodder-shelter is preferred. This ensures that the building of sheep stables entails not only cleanliness, adequate ventilation, and other conveniences, but also, in certain areas, beauty, if not luxury. It shows that sheep breeding has advanced significantly. For instance, I saw a sheep pen destined for lambing with not only windows, but also window shades, in the belief that too much light would be harmful to the lambing mothers. This, I think, is a group of sheep owners who are willing to part with 100 Florins[5] for this. In general, since this year was a good financial year and the value of wool increased, the construction of these pens, especially for lambing, has increased in many regions. **Administrator Rudolph André's classic work on wool enhancement is well-known, and his ideas have been widely adopted and implemented.** Individual pairing has been observed in a number of herds that have significantly increased. **The owners can be seen inspecting the breed registers themselves.** I know a few people who were excited to start cognate herds and invested a lot of money on rams and brood sheep. Both well-known agricultural institutions will claim to raise sheep for cognate herds. I also know of certain sheep enthusiasts who consider themselves particularly fortunate after purchasing imported, authentic Spanish sheep with provenance records and introducing them to their herds. Not only in my field, but even in the Tisza, Maros, and Siebenbuergen[6] Frontier regions on the other side of the Danube, where there were once none, or only a few Zackel sheep with bad wool, one can now find fine flocks of a thousand animals with reasonably improved wool.

The following is my response to the question of whether these owners and their staff have sufficient knowledge of the characteristics of wool and whether their purchases are worthwhile. These gentlemen who purchase rams to increase a herd of 1000 sheep will receive one or two outstanding rams, as well as a few poor rams that will not harm their flock. They also purchase animals from various herds in order to experiment and develop a method for potential work. Those gentlemen who purchased breeds should find themselves in trouble if they do not carefully choose the characteristics. Buying from a variety of farms, it seems to me, will be the best option for avoiding trouble. **They have true breeding registers that can be consulted if one is unsure which institution has a certain animal with a specific form of wool. The future will reveal whether they pursue inbreeding or attempt to minimize any negative traits by cross-breeding.** What I saw was a mixed lot in which it became clear that animal owners did not necessarily have the right option. Officials are often fooled by events and personal interests. **I would like to add my humble opinion at this stage.** Since returning from the Association's meeting in Brno, I was met with skepticism because I stood firm in my belief that in Brno, the prize was not given to the best animal, that no license was given out, and that the animal could not be purchased on the spot, but that people like Geisslern, Lamberg, and others seldom or never sold the animals they brought to the market—one had to visit their farms and obtain them there.

When the merits of our Association attracted sheep breeders from far-away regions to Brno—when the debate kindled farmers' interest—when scientific papers, the manner in which negotiations were conducted, the gathering of so many distinguished wool producers, wool traders, and wool manufacturers, and even the commission of judges—when all of this instilled confidence in the less educated wool producers, traders, and manufacturers—when all of this instilled confidence in the less educated wool producers, traders, vanity and the feeling that they knew best did not lead them astray, according to what I observed in everyday interaction with busy wool producers. They all wished, though, that the animals that won prizes in Brno be sold there as well, preferably with a license. It will be a milestone in the science of refinement, that in the year 1819 a mathematical definition and naming of the grades of wool fineness was established. But, according to my insights, a few other characteristics remain to be mathematically ordered: the smoothness, the elasticity, and the resilience in equal fineness of prime wool.* The majority can be seen just by looking at them, though they seem to be more a matter of personal interest. The density and softness of the fleece are among the first; the white fat and the length of the wool are among the latter.

What prevents the Association from selling animals under its license at sheep trade shows in the future if the traits are so distinct that they can be named with a number or a term that dynamically reflects the concept of the incremental character? This means that all livestock sold by this establishment would be assessed by a sworn commission before being given to the new owner with a certificate signed by the licensed gentlemen. Will this mean that the Association will only license ideal animals, such as class 6 animals, or will it still accept animals from several other classes? They will race to follow the trend of improving the skillfully driven and costly pursuit to excellence, where they will be rewarded. But it is also important (and necessary) to avoid investments that will cost them not only time and resources, but also their enthusiasm for a worthwhile endeavor. Could I make a rather heartfelt proposal on my own behalf, but without any personal vanity? I would like to display some of my own animals at a forthcoming exhibition of this Association that has been so helpful to me, and the proceeds from the auction of the first of my registered rams will be donated to the Franzens-Museum.

Patty, October 15th, 1819
Your very respectful servant, Count Emmerich
Festetics as Representative of the Eisenburg Committee

---

* I do not know the private goals of the excellent sheep breeders, but I doubt that they would trick purchasers into buying their animals for more than they are worth. They all realize that it is to their advantage to dignify the reputation of the acknowledged authorities. C. C. André.

1820. *Economic News and Announcements* **15(20):115–119**

2. Remarks by Count Festetics

The undersigned had the distinction of submitting his reflections to the widely esteemed Association **on the observations, scientific denomination, and eventually the enumeration of all the qualities of a great wool, beginning with fineness and progressing from there.** The valued Association initially advised me to wait for Counselor Thaer's view. Counselor Thaer's remarks were written in Nro. 12 and 13 of the 19th Edition of the *Economic News and Announcements*. It is my intention to advance this important branch of agronomical science as much as possible that I hasten to make my individual statements concerning this important article, as well as the remarks of Secretary and Advisor André. I am keeping track of the publication's figures. First, it is true that the Association would have to be mindful of the descriptive terms used by wool merchants and companies and that would have become part of the working vocabulary. And more so, given that the Association should take the lead in naming these characteristics. Counselor Thaer said categorically that a wool trader told him the wool had the characteristics of "electoral wool," despite the fact that it was not "electoral." However, I believe that the Association should not be too undisturbed. We have examples of scientific societies instructing the common man, without bias and prepared with experience, that **the Association, incomparable in its knowledge of the science of wool, had the ability to discover and call the characteristics, to compile an aggregate of their lesser or greater excellence, and to have justification thereof.** As a result, the wool trader will be told, and our manufacturers will be able to articulate their jobs, while before they could only speculate without getting a firm idea. The manufacturers will be able to remember the features of the fabric, either individually or collectively, and the Association will be pleased to have been able to organize and clarify the wool traders' ideas.

Also, to have scientifically aided the manufacturers, and to have given the sheep growers a clear set of rules. Counselor Thaer's foresight and scientifically direct vocabulary, as well as his paternal involvement in our efforts, have greatly aided this effort. I can clarify that I am not about the producer's conditions; even if it is true that not all wool can be used for every commodity and that the producer might be in a position where he is reliant on the manufacturer. **The point here is to determine which characteristics the perfect wool should have. Do these characteristics have gradations among themselves? How can one name each characteristic, each in itself according to their mathematical gradation, using a scientific terminology? Finally, three questions are the most important: Can one find these characteristics in their highest (as far as they occur singly in Nature) form in combination? Or are some of the characteristics—and which ones—where one will exclude the other? The second question is, if the characteristics in their highest level cannot be found together, should one then segregate between wool characteristic and characteristic,** similar to two different wools such as Escorial and Negretti, and conduct further examinations? Or does one conclude that there is no such thing as great wool, and therefore Escorial? In this case, the Association will be justified in determining the most pleasing combination of characteristics. Following that, it would be determined which wool style production would be the most beneficial to the farmer's financial situation.

The third, fourth, fifth, and sixth questions are as follows: I don't think the Counselor here meant to imply that a well-made, superfine wool would not please a fine-textile manufacturer. The third question seems to me to only suggest that a chosen portion of a mixture of fine wools would, for example, be the most useful for combed wool called Electra, rather than the features of the wool itself. The fourth issue suggests nothing more than that the finest wool may be classified third class due to a lack of required characteristics, even though the thickness of the wool string would rank it among the finest.

In Hungary, we have long known that the wool with the least thick fleece are those with the smoothest appearance; that the open tips of the wool rise under a constant shortage of fat as soon as the sheep is exposed to air and rain; and that, in the end, they lose their toughness to the point that they snap and tear like they have been half-singed. **I believe that the knowledgeable remarks of Mr. André are the object of further work which the Association will have to consider with regard to the naming of the rest of the characteristics.** However, I will immediately remind my most esteemed friend of my observations in relation to question Nro. 6. The author of these findings, G., unequivocally notes that the wool tips of certain animals with darker, steadily turning to black, resin-like fleece have lost all useful characteristics due to that rough crust. That is something I can admit. These wool tips lose some of their perfection when they dry out and the wool becomes brittle, not because of the accumulation of fat there. However, the advantage to the rest of the wool is significant because it helps to make the fleece denser, allowing fat to accumulate in the rest of the wool and preventing dust and sunlight penetration. This trait is almost necessary and *sine qua non*[7] for nomadic sheep, such as ours, who must most of the time find their own food. They are held next to the house on the estancias, in smaller herds, and, like us, in outstanding dairy farms. They have a less thick fleece and lack the crust, allowing them to stay under the Sun's influence. However, I often observed that the wool of well-fed stabled sheep included 3 parts clean wool near the body, 2 parts dirty wool, and dust-free yet fat-free outer ends. I am familiar with one herd that has real Electoral wool and is also known to the Association. It is continually exposed to fog, wind, and heat, and as a result, the tips of the wool have lost their toughness and feel singed to the fingers.*

This comment will cause me to suspect things later. All else is based on my strong convictions, but it is not related to the study. And seventh and eighth I absolutely approve. Despite my aversion to interpreting Counselor Thaer's inference, I believe my friend André went too far. If the result of examples #1 and #2 is full Electra, then the other observed features, apart from fineness, contributed to the classification.

Furthermore, I believe that the Counselor did not want two levels higher, and that only two levels of the following wools would be used as the fineness gradation measuring norm. The especially quoted words of the English wool merchant, on the other hand, were an indicator that there is a select one, an Electra, in each characteristic of the wool. That an Electra will have to be worked out even in Electoral wool. He went on to say that here aside from the

---

* Was this not the knowledge which moved those masters, the exemplary sheep breeders, to erect shelters on the grazing meadows where the chosen found shade or a dry place during rain. Others gave their sheep clothing to protect them and fed them in their barns. C. C. André.

high grade of fineness, also other characteristics play a role, without which the highest fine flakes would be thrown into the third. **Since the Official Rudolph André's micrometer measurements put all Saxonian wool samples in the class 5r55, the example#1 .zzs, and #2 .gr*55, then it would appear as if all Saxonian wool would be excluded from the Electra qualification.**[8] This, *de facto*, would mean that we would have lost our process with all the wool traders and manufacturers.

There is no answer to #9. The Counselor explains himself objectively and simply in response to #10. I have come to the following conclusion: real Merino or Spanish sheep may have a thick fleece of fine wool, full of muscle, and thus feel rough to the touch. The wool is so thick and curled here that it seems short, even though it is the correct length when extended. Despite the fact that this concept is rather abstract, the Counselor called this particular wool trait Negretti. Real Merino or Spanish sheep, on the other hand, may have similarly fine wool without displaying a thick fleece that seems longer and has a special soft touch—this will be Escurial to a high degree. I comply with Mr. André's observations, but in the case of the last, I would have preferred that the traits not be reduced to fat and fitness temperament. This is a good time to point out that the Association has a lot of justification for believing that these two extremes should have their own names, and that each race should be treated differently because of its own unique characteristics. **This differentiation would be the desired progress in the higher science of sheep breeding. With this, the sheep breeder could set himself a goal according to his individual position, and he could reach it in a secure way. If the Association creates a new classification from the cross-breeding of these two classes and those new characteristics, and how that should be treated, stands for future discussion, when the characteristics of the wool from this new herd show constant and dominant.**

To #11, the whole paragraph is critical, and it has been debated many times for the same outcome. In another place, I heard a discussion about whether the term Electa refers to a different wool fineness classification.

To #12, the first few sentences in this paragraph contain the most critical sentences. I will give you a sample of what I am talking about: The exceptionally fine Escorial wool is used to make fine textiles, for starters. Negretti wool, on the other hand, is used for a more durable fabric. Second, the Escorial wool-bearing sheep are a particularly fragile breed, according to the Counselor's experience. Third, the formation and survival of such a herd entails many challenges, which would obstruct the herd's expansion in and of itself. To the fourth point, the ideal length of wool has drawbacks, first because the wool does not stay straight, but rather lays down. And, since these animals are often frail and sickly, they require special attention and grain rations. **Fifth: there are factions where it is said that this affliction can be cured by mixing blood. Together with Mr. André, I want to exclaim: So many points of importance! Sparks and flames erupted when one accompanies these paragraphs with observations.**

The first sentence is mine, from an essay about inbreeding and has been discussed. Since Counselor Thaer now repeats it, I would like to remind the Breeders' Association that I hope to finish the work of wool classification earlier, if we consider two, or even three races, such as Negretti, Quadeloupe, and Escorial, we can see how similarly fine (say—superfine) wool can have very different characteristics.

I have a feeling about the second sentence, which seems to be backed up by local knowledge. The so-called Saxonian wool is identical to that produced in Spain, where conscientious Spaniards keep their sheep on grazing land designated by statute. In the stalls, selected Merino animals were housed and mass fed, equivalent to the Moravian half meal. The rams were chosen based on how fine the wool fleece was and how shiny it was. We focused on these characteristics in the selection of rams in our area as well. **In the third and fourth generations, we found that we were getting less wool than normal.** Negligent handling resulted in bare bellies, with the most severe consequence being the rejection of the whole fleece of underfed sheep. **But if the gentlemen would stick to the strict truth with the recommendations of refreshing the blood, then the Association should investigate, whether they would either want to sell their rams, or whether they, because of the above circumstances, had to rely on the blood refreshment themselves and then recommended it to everybody else.** This Escorial wool, if it still retains resilience and strength, will become necessary for newly discovered products. Therefore, a constant and extremely careful tending to each animal is highly recommended. But, since **the Austrian State cannot afford such tender care for millions of sheep,** the breed Negretti and partially also Quadeloupe will make up the majority of the sheep population. Now to Mr. Neverent's question, whether it was still advantageous to breed this kind of sheep, since the finest of wool examples from the herd of Count Salm had been eliminated. Here one has to ask oneself, is the diligent overseer, **Mr. Rudolph André, who deserves warmest thanks from all sheep breeders, able to elevate the entire establishment to this high standard of fineness? Is this hope based on the high noble parents and ancestors? Or does he hope to make this anomaly among the herd permanent?** I ask: will last year's yearling little mother give the same z**ss wool in her second year? Did they not spin any wool despite their usual careful tending? In all this lies probably the answer to Nro. 3: that a very extensive acceptance of the Merino will always be impossible, because of its expensive maintenance, and the reduced weight of its wool.

The truth of #4, as far as the damage of the skimpy wool is concerned, has been felt for a long time. In connection with the statement that the fineness of the wool is dependent upon a weakness or disease, **I would like to cite my 28 years of experience.** I saw that in the improved herd, which I tended in a better, but rural manner, the mortality was highest among those that had been improved per saltum. I am asking all sheep breeders, who used the most noble rams to improve their herd, whether they did not bemoan the mortality of the young offspring, that the most beautiful, that is to say, the finest wool bearers had died. **I cannot and will not differentiate whether the fineness of the wool prompted the weakness, or the improvement of the already weak mother sheep caused the birth of a weak lamb that had inherited the characteristics of the ram.** I did not find that a true Negretti with very fine Escorial mothers would result in better, uniform wool than when an Escorial ram is put together with a Negretti mother. **But in my inbreeding, I always choose rams which had very fine, relaxed but strong, and dense wool fleece. I learned from my experience, that the apparent fineness was tied to a loose fleece,** and from that sprang a floppy, hanging, spinning, and light wool. Rudolph André assured me, that a Mannersdorfer or Theresienfelder ram, who even has prickly wool, would improve the herd greatly, and he therewith admitted that this ram with the dense fleece, crossed with a Paduaner sheep, would bring the best, almost too much to hope for, improvement.

This is also an answer to #5 - ad 14th. Counselor Thaer specifically wanted that here we keep the differentiation between the true original Merinos. **This I see as the only way to clarify our terminological attempts to the manufacturers, wool traders, and producers. And also make it the crowning of our work.** To call them Negretti and Escorial, I think, causes less confusion than if we were to use the word Electra just because of the one characteristic. Whatever the perfect Negretti, or the perfect Escorial will be, and which still to be named characteristics belong to each, will be the next easier task.

Patty the 30th of March, 1820
The Association's most sincere Servant
Count Emmerich Festetics, Member and Representative.

1822. *Economic News and Announcements* 92:729–731

373. About an essay by Mr. J. R. in the
third publication of the year 1821S

Apart from other business activities, the practical aspects in some of the agricultural branches take so much of my time that (even if scientific knowledge could come to my help in all areas) I could not possibly proceed to give my judgment on all three papers. My visits to the Sheep Breeders' Society in Brno show that, although I rarely engage in debates, I have always sought to improve myself, to refine my ideas; I have also always enjoyed being corrected. However, I see that I am listed on page 151, first column, and that opinions are solely focused on observations I made an on my so-called "bogus logic" while researching Hungarian poultry farming. As a result, I am very concerned that this, like the rest of the article, could lead to extremely fabricated results with possibly disastrous implications for any sheep breeder, so I see that **I must once again take up my pen to write on the much-discussed topic of inbreeding.** First, a few words to the paper's editor, Mr. J. R.: You Sir, seem to have read my papers on sheep inbreeding from the bottom up, more in the style of an all-important and very busy gentlemen.

You saw a count advocating inbreeding, read backward the arguments, and by chance [*Econ. News Ann.* No.92, 1822] discovered what I had scarcely touched on at the outset, namely the newly populated house Henne and her brood. You may then have set the text aside without reading any more, and according to your article, **you believe that breeders in Hungary are attempting to turn ordinary hens' feathers into the plumage of humming birds by artificial mating.**—No way, Sir! **The Hungarians are well aware that their animals, which are already reliant on special breeding compounds for treatment, must strictly adhere to Nature's breeding rules unless the same nature, which is otherwise so forgiving, harshly punishes deviant conduct.** Finally, there is the Sheep Breeding Committee! My essays had to keep up with the current controversies. **In his useful article, Baron von Ehrenfels questioned interbreeding. The committee chief, Baron von Bartenstein, who wanted every single relevant sentence clarified in detail, added the following topic into the discussion: whether inbreeding is superior to the historically**

recommended breeding form for a given livestock. The below are my interpretations of the following words: **I define breeding livestock as any species that retains primarily the main traits sought by the breeder, those very characteristics that they themselves inherited and are likely to transmit further by careful breeding: a constant must be the pairing of desirable and complementary characteristics such that their combined undesirable traits are eliminated. However, I refer to the process of grafting as the addition of a homogeneous animal's unique feature to the stock's many mastered qualities.** So, if you are still fortunate enough to have a small herd of breeding livestock that **can still bequeath and inherit attractive traits, you can avoid interbreeding and stick to (general) inbreeding instead. The topic of inbreeding was the focus of the discussion, and Advisor André, secretary, selected this moment to document it in full and exhaustive detail on the pages of "*Econ. News Ann.*".** However, he also used the terms "necessarily" and "long-term" next to consanguinity. **In none of my essays did I ever state that mating should be limited to the next consanguineous relationship;** rather, if the designated breeding ram meets my ideals, he should be used without regard for consanguineous relationships.

There was only one ram for the nine mothers I used to establish my lineage, so its lambs were genuinely brothers on the father's side but more distantly related on the mothers' side. There was never more than one breeding ram during the 16 years of multiplying moms, resulting in an inbred condition. The name, however, was not really my most important law. The most important thing was for the ram to excel, for the sake of the breeder's prestige, in outperforming the competition in terms of fleece consistency and organic construction. Should the fact that my definition of organic strength is a well-fattened sheep be used against me, I can only claim that I feed my sheep neither oats nor peas, and that my livestock is always kept a safe distance from the maintenance place. **I have never shared an opinion about how long this inbreeding should be allowed to continue, as it makes preventing blood relatedness difficult.** For the first time, I am stating that because supervised inbreeding has been done for so long that breeding mothers must be faced with multiple rams, it is time to divide the herds into families and impact each with a different ram. And if there are now households, this does not warrant selecting a ram that tends to depart from the original or ideal. **Even stringent**

inbreeding cannot deter variations or new forms brought about
by the often-active natural world, as all observant sheep breeders have learnt from experience—and will not doubt—that there
is no breeding livestock on the Earth in which variations or new
forms brought about by the always active natural environment
can be avoided. These variations can be found on individual body
parts or in the wool's properties: Finally, the wool has much bigger
or lesser properties. If we now introduce a ram to one of the families that deviates from the ideal by a small—yet fun—discrepancy
and allow mating to occur, these marks will continue to diverge
with each generation. Thaer, the town counselor, recognized this
right away. Dr. and Prof. Fuß interprets those crucial terms differently than I do: namely, that when it comes to mating, if one's
emphasis is solely based on the highest grade of fineness and softness (which is very obvious) **and one sticks to a single top line
of heredity, minor shortcomings in density and compactness
result.** It follows that, if one continues to search solely for continuously higher fineness while ignoring all other criteria—this after
I have battled those terms in all my papers—all other suggested
values must be similarly included in the most polished wool.

I must reiterate, particularly when considering, of course,
supervised breeding, that the ram's wool, in comparison to that
of the mothers, must maintain a certain virility. I also took advantage of the chance at the Brno fair to point out to my neighbors
how, despite having finer wool than the mother, the year's twins,
born from father and mother simultaneously, have characteristics
in their wool that expose the mother's sex, something I have seen
many times among my herds' regular yearly twins. Mr. Rudolph
André, I think you have grasped my meaning. I still believe that
cross-breeding definite Negretti breeders with true Escurial
breeders can only yield bastards; that it is more difficult to maintain the combined features of each in the next generation than to
see them revert to one or the other of their original traits.

I also claim that, with our great care, careful selection of rams,
and feeding methods tied to a minimum expanse of movement,
**there will emerge a predominant tendency for lighter wool vari-
eties to arise out of the original dense wool strains;** this choice,
of course, means trouble for the sheep breeder; I would even go
so far as to say that he might also need to overcome his own resistance. Isn't that argument supported by anything one has ever
heard about the physical inspection of the lambs that the Majorals

in their Lavagnen are inspecting? What Dr. Fuß says about characters on page 66 may be considered as a harsh critique. In this regard, I would like to direct the individual to an article in the *Economic News* that explains the ongoing race in Koptsan but only provides a few but comprehensive details regarding stallion selection. On the same article, the handling of Hungarian farmyard animals is discussed; I agree with half of it, but I would clarify that bulls are still purchased and taken into certain herds from the outside, either because the stock was never mastered or for other purposes. However, no rational breeder insists that the bull alone will make the difference in the absence of improved supervision and treatment. But where do the Deregegyházer, Pahier, and Lutinovitoische Gulya get their bulls, from which every single beautiful breed is purchased? They raise them themselves, and the Tulipán breed in Deregegyháza has never been exceeded as the result of their inbreeding, never been surpassed, and the same is true of the breed from the Einsiedel monastery in Switzerland. **When Bakewell, Buffon, and Sebright performed experiments that went beyond the laws of Nature or else attempted to force against it, the all-powerful one, a model that lies outside of Nature's bounds, they are no less deserving of our thankfulness, because we learned then that when methods are used that go against Nature or natural processes, Nature will retaliate.**

Therefore my conviction holds—and along with the founder of the Agricultural Scientific Society, the honorable town counselor Thaer, I am saying that **conceiving within the closest blood lines is generally not harmful and that, much to the contrary, it ought to be practiced if the intention is to keep reproducing the highly valuable and desired qualities previously achieved in individuals and to make them permanent in the stock;** and this is still true for someone lucky enough to own a breeding stock nursery. I want to close in Dr. Fuß's words: namely how, if the stupid, shortsighted—lazy, ignorant, careless reader, any rich person, etc. were already assured of success while working under all circumstances and conditions through blind imitation, there would reign everywhere the blind luck of fortune—whereas reflection, sustained attention, clever perseverance, God given gift, intelligence would never find true reward, not even satisfaction.

Güns, January 1822. Count Emmerich Festetits

Appendix 3—Explanations of Christian Carl and Rudolph André

1819. *Economic News and Announcements* 4:26–27

Debate: Sheep Raising

Explanation of Count Emmerich von Festetics
(Conclusion of Supplement No. 3)

Remarks of the Publisher

I think myself completely rewarded for the utmost honor and kindest reward (suddenly to me for a multitude of distorted judgments of superficial knowledge, whereof have been understood by me only once), that such a character, mind, and origin of a man, such a Hungarian veteran in higher sheep breeding, like Count von Emmerich Festetics; my opinion of the explanation thus instructive, well-founded and natural, has deemed worthy of supported discussions. For my part I find nothing essential, in which I must not fully agree. **The count himself recognizes the necessary modification of bequeathment of those closely related by blood, that it has not occurred unconditionally.** In any event, I would nevertheless still want to give slight importance which the claim mentions: **to avoid the consanguinity as the rule. This is freely a more subtle, perhaps more hypothetical point, our natural philosophers have made here, of my judgment, a very lucky brood, and I would appeal briefly to half of them, when it would be only accommodating to them,** and discussing these clearly and intelligibly. I want to carry out a risky experiment, to point out the meaning of my opinion in a few words.

It seems to me a law of Nature, that not the homogeneous, but rather the heterogeneous by reciprocal important reaction of the stronger operation, brings forth new products, in shapes and forms. Otherwise, in what does sexual difference consist, then? Are both sexes not already a heterogeneous opposite? In addition, however, there was contradictory to this principal, not, yet a multitude of subordinates, the individuals are not at all without influence, as very few are taken together? This is clear to us with people, because we observe them more distinctly, more sufficiently, according to their differences by many nuances. It is scarcely to be doubted, for example, that when blue-eyed blonds throughout marry in several generations the most possible closest relationship: still turn out to exhibit weaker constitution in the latter descendants, than the blond variety in general already separately tends to be. And that is only the first homogeneity, and how many of them are there still to be observed! There are princes and other noble families, where this closely observed bond indeed expressed a striking *air de famille*,[9] but not directly to the benefit of descendants, whereat debilitation cannot be mistaken. Perhaps this example explains my meaning better than a long deduction. With horses one has likewise employed enough observations and taken into consideration subtle qualities in pairing, at one's command, not allowing covering mares in close relationships, certainly, for example, not a phlegmatic brother with a phlegmatic sister. In the case of sheep, there were only very few owners, earlier still shepherds, who thus knew the individual nature of each single head in a herd of 600, that they promptly knew to discover and to characterize: There are too many animals and the apparent dispositions too great, and still yet **there is enough observation to find here homogeneity and heterogeneity of several kinds;** whatever attention gains and qualifies it, whenever instead of my giving the sister to a brother, I prefer a much more distant relative from prevailing grounds.

Should not some owners of herds have made the observation, that certain rams possess little or no procreative power? Does this not become similar to a critical basis; is abandoning in such a case consanguinity decided beforehand; but rather does it also raise the question entirely: **from what place this weakening originates? Allowed to be found perhaps the in the family genealogy? And how many homogeneities and heterogeneities does the nature of the wool present? Does the so-called renewing of blood have further its own sense and basis? Or is it a fantasy? Analogy occurs according to natural law in the plant kingdom! Seeds from the same stem degenerate finally, and does the growth of seed have a good basis or not? The more freely nature is allowed** remain, the more similar and truer it remains open with more certain strength in its formations—descendants outside for a long time. However, so how will the position of Nature work in the case of civilization among humans, abandoned in connection with home discipline among animals, weakening and disposition toward susceptibility and sensitivity did not begin there *eo ipso*[10] against deviating influences, detrimental to full health of the organism? Still enough for an indication, but so much for little instead of fundamental exhaustion, which the masters will give us.

However, I must pay respects to the Count by my wholly special gratitude, that the first is, often and freely admitted, which about the condition of our sheep breeding and our handling of wool, perhaps unexpected and not acceptable in *Hesperus* (January, 1818), still in the purest patriotism and inner reality—cheerfully finally once it made ring in some ears even thus paradoxically, believed to be compelled to say, and that value lies afterward, that I indeed wish to him; however, I was impossibly able to express myself. Nevertheless, I believe to have exposed some truth in that article, who composed this and that, still nowhere said in this short space of time, and wanted to be done clearly from little, and who quite understand, and our sheep raiser considers in the shortest time and most certainly those (who already are not decided about the masters and their way) would be obliged to lead toward the desired aims that our friend, who deceives by flattery, or that one, says to support us, how does he find us?

Christian Carl André

1819. *Economic News and Announcements*
21(Supplement):161–162

My Views and Remarks concerning Organic
Weakness, especially in Sheep with Fine Wool.

In response to Count Emmerich von Festetics's
Essay in the January Issue 1819, Insert No. 2 and No. 22.

Generally, I will agree with the respectful Count's declarations
in the January Issue, concerning organic weakening; however,
I shall herewith attempt to present my views in a generally
understandable manner. **I am, however, going even further
than the Count, in that I consider inbreeding as the main
way to defeat the organic weakness.** My proof is the following. **Organic strength of a fully developed organism can only
be found in an animal that was paired with an animal of the
opposite gender, but with the same characteristics, and which
then produces offspring showing the same traits.** This can
only happen with purebred sheep, and therefore it is my opinion that **purebred animals are the essence of organic strength.
Homogenous animals that breed among themselves generally
produce strong organisms in return.** One could for instance
ruin a purebred flock by misinformed treatment. However, if
then left to pairing among themselves, I expect them to recover
their former state of well-being with the help of good treatment
and inbreeding during several generations. They will once
again produce wool of the same quality and quantity as before.
And the pure fact that one can achieve this is proof that a purebred animal generally always has organic strength—or better
said organic resilience. **If this assumption is correct, then I
could make the characteristics of bastards or mixed-breed
animals permanent by inbreeding them and therewith create
new races.** But no breed is viable without organic strength and
firmness, and therefore inbreeding is the best way to prevent

organic weakness and, in the long run, to assure a solid organism among the flock.*

To the definition of organic strength of the purebred and the organic weakness of the mixed breed, I would like to add the word "permanent" to make it clear that a mixed-bred animal will have the organic weakness for as long as it lives, just as the purebred will have a lifelong organic strength, as long as it is in good health and gets good treatment. Sickness and bad treatment will cause organic weakness, which however is temporal. And with these observations, I cannot think of the quality of the wool as a product of organic weakness, I see it as I see the fine conformation of an Arabic stallion.

Raitz, March 1819—Rudolph André—Administrator

---

* Generally, organic weakness can only be found among bastards or mixed-breeds, which are the result of two distinct organisms that were blended— that is to say, the father was one, the mother another—so that the result was a new, not fully developed organism. The final development only occurs when these mixed-breed animals inbreed for several generations and when, for this purpose, only animals were chosen that presented the strongest characteristics of the mixed-breed ancestors. Then the characteristics become firmly established. **The resulting young will then eventually present a firmly established organism, which will then be reproduced** *ad infinitum*.[11] **And in this way, bastards will then be elevated to purebred status.** This is the way to solid breed improvement. C. C. André.

Appendix 4—Festetics's Letter to Hugo Salm

My Dear Count,
Honorable Director,

Unfortunately, your letter of December 31, 1820 was delivered to me only on January 22 of the current year, hence I strive to reply to your benevolent writing as shortly as possible.

You say that there remains an obligation to be fulfilled for you as Director of the Agricultural Society, namely to suggest that the merits of our resigning Secretary André should be acknowledged through a rightful official post of national importance within the Austrian realm, and thus you amicably request me, a member of the Society, to pronounce, according to my conscience and conviction, whatever I can put forward about André's merits.

Sir, your request puts me into a most embarrassing situation. How could I give a general overview of the fact that, for a good many year, André has been acting upon his best intention and practice, without official employment and related compensation or gratuity, having chosen, as a foreigner, Moravia as his new homeland, mostly and necessarily adapting to the interest of this country and the nature of his efforts, insisting on the advancement of neighboring Hungary too, with unrelenting vigor, advocacy, sacrifice and increasing enthusiasm?

Before I could propose a short acknowledgement faithfully, let me remind you that, for my own part, having been busily engaged in agriculture and led by the star of the wise to Brno, I am now actively working in the same benign Moravian and Silesian Society, encouraged by André, known for the great knowledge and clarity of his writings, which myself, with sincere friendship, received and presented to those gentlemen who, having understood his intent, provided protection for his efforts, to my great happiness, while he was also able to please this circle of laboring men with moments of compassionate bliss.

I was entirely exhilarated by the discussions that I could attend so far. There remained nothing to wonder about the fact that such a great number of Moravian officials, to equal extent and significance, received the training, by due ambition, that earned them, to the highest degree, the attention of the most rigorous censors.

*As far as you, sir, as a Director as well as the fellow presidents of the Society generously acknowledge now that André can be described as a founder in creating and setting up this Society due to the outlined plan and his tireless efforts, that he occasionally also restored sufficient efforts on the part of other members relying on his own endeavor, even if he had to risk his personal standing and health in the struggle for the good, it cannot pose a heavy sacrifice for me either to admit that I began to develop my own agricultural intelligence since **I learnt to organize my views in this association. Since then, I do not believe that there is anything better I could do in my surroundings than having the courage, as a landowner and an official, to visit the Olympus of agriculture, and through this effort and the growth of knowledge I became more efficient and more enlightened here, at home, where these two good guides aided me in their dissemination.***

Since then, I feel and did proclaim that I am obliged to express my gratitude; I also admit that I have certainly frequented our general meetings for two years with the idea that I should voice a proposal which you, sir, now follow as a true patriot one thing, however, prevents me to do so: apparently, I am afraid that Hungary intends to bring considerable reproach on this Moravian sire.

Therefore, I can, following the most sincere dictates of my conscience, explain that an award initiated at a higher level, one that meets our expectation and is also acceptable for Mr. André, our departing Secretary, could be a benign act of the government. Such a gesture would address Moravia, for whom this fine man worked primarily; moreover, it can not only reach out to the agriculturist audience of this entire realm but to the universe as a whole, while they could gain a supporter of the agrarian domain who is able to transcend his own selfish interests and stands before us with all his merits demanding our gratitude.

What is more, such an act would represent sentiment and goodwill because it could support this worthy man and the subject of his works, winning him over and making his works effective and esteemed at the same time. This gesture would also provide an opportunity for delightful prospects because it could raise hope for our aspiration that this society can experience its existence through its very members. Besides, it would also encourage André's descendants to make similarly tireless and unselfish efforts.

With most faithful friendship and best wishes for you

Kőszeg, January 23rd, 1821

Most obliged servant and representative of the Agriculture Society.

Some signatories, living nearby and signing as members of the Society, hereby declare that they fully agree with the statements of both Count Festetics and Count Salm, and express their most sincere gratitude with good wishes for your most fortunate success.

Ferenc Chernel, assessor, Imperial and Royal Regional Court of Appeal

György Chernel, manu propria, member of the Moravian and Silesian Agriculture Society

József Farkas, bailiff, member of the Moravian and Silesian Agriculture Society

Appendix 5—J. K. Nestler's Lecture About Procreation and Heredity

1829. *Announcements of the Imperial-Royal Moravian-Silesian Society for the Promotion of Agriculture, Natural and Regional Studies in Brno*

47: 369–372; 48: 377–380; 50: 394–398; 51: 401–404

The Effect of Generation on the Characteristics of the Progeny

Publisher foreword

Notwithstanding the very praiseworthy achievements of individual excellent animal breeders, it is well known that the principles of a rational procedure in the various branches of agricultural animal breeding have not yet reached that degree of transparency, independence, and generality that would be obviously desirable for their successful prosperity and would be able to guarantee it with certainty.

Accordingly, it is our pleasure to present this treatise, which is a part of the exemplary lectures of Dr. Nestler, the Professor of Agriculture and Natural History at the Imperial and Royal University of Olomouc, where, as early as in 1827, the meritorious author motivates the most important aspects of rational animal breeding in their various relationships, and develops the most important procreation influence on the characteristics of the offspring with a critical spirit, in a manner appropriate to the present state of science and practical experience that certainly deserves the general gratitude of practical farmers, and, thus, at the same time, made a valuable contribution to the thorough solution of the issue raised both at the meeting of the Sheep Breeders Association in Brno this May regarding the Merino breeding principles of Baron von Ehrenfels, and that has been dealt with in No. 10, 17, 18, 19, 28, 29, 33, 34, and 45 on these pages ever since.

§ 1

Procreation is the most important, and often the only possible mode of reproduction for most plants and animals in the agricultural household. The product of fertile procreation is the offspring more or less similar in their characteristics to the ones they descend from, depending on whether the parent passed the essential characteristics only or also the accidental characteristics to the offspring.

§ 2

A fertile reproduction, where the offspring inherits all essential characteristics, is only possible between two genera that belong to one and the same species (in the natural-historical sense).

For example, between two genera of a sheep, cow, horse, etc.

§ 3

A fertile reproduction with loss of some essential characteristics is, however, possible as well, under special circumstances, between two genera that belong to one and the same genus, but to different species. The offspring of such a fertile reproduction is known as a bastard, and the mating itself—a cross breeding.

Thus, the family of horses mates with donkeys and zebras, while the dog family mates with those of the fox and wolf. The male goat jumps on the mother sheep (what I have seen myself many times). The pheasant successfully mates with a house hen. Hence, we have bastards between the canary females and the males of the siskin, goldfinch, bullfinch.

Crossbreeding between related species of a genus is even more frequent in flora, as we experience in flower and kitchen gardening, where crossbreeding is mediated by drafts of air or wind, insects, etc. by transferring pollination dust to the fertile flower stigma, or is deliberately initiated by plant lovers aiming to obtain new modifications.

§ 4

In case of crossbreeding between the genera of different species, the one can clearly see,

   a. that the characteristics of both genera merge in the offspring;

b. that the mother body, which nourishes the generated germ with its juices up to a certain stage of formation, in most cases transfers more of its characteristics to the offspring than the father body;

c. that the bastard reproductive capacity in fauna is almost completely extinguished;

d. that the bastard fertility in flora is, however, preserved in the vast majority of cases;

e. that, for causes not easily explained, there are sometimes exceptions in terms of the similarity of the progeny to the progenitors in the inheritance of the reproductive capacity.

To a) We find the characteristics of both progenitors fused in the mule, as the offspring of a male donkey and a female horse, same applies to the hinny being the offspring of a male horse and a female donkey.

To b) The mule has the size, the voice, mostly also the color of its horse mother, and, in general, is more similar to her than the hinny, which reminds in size, voice, color again far more of its mother (the female donkey) than of its father.

To c) Mule and hinny are usually infertile. There are very few cases of real fertility when the offspring was actually carried out and born in due time. These cases are considered very rear. Mostly bastard mother animals abort, or their new-borns die soon after birth.

To d) The kitchen and flower plants bastardized by seeds are usually capable of germination. Köhlreuter continuously used intentional crossbreeding to transform Nicotiana rustica into the red-flowered and quite differently shaped cultivated tobacco (Nicotiana tabacum) that he then returned to Nicotiana rustica with continued crossbreeding at the end.

However, I have come across a bastard of a spruce and a lark tree on the road near Türnitz in Austria, which was completely fruitless despite its free location favoring the seed production.

To e) In Spain, where the mules are more protected and three, four times more expensive than horses, the mule value is the higher, the better characteristics of the mule it combines, namely great endurance in strains, strong health and good thriving with poor care and feed as well as its beautiful forms of the horse.

§ 5

As we notice to a greater degree in the offspring of crossbreeding, similar phenomena can also be found (to various extent) in the offspring of two genera belonging to one and the same species, because

a. the characteristics of both parents merge in the offspring.

   For example, if the completely living in the hot zone has simple fine hair; the sheep of the temperate zone has brown-white, whereas mating with the black cow will bring black and white offspring.

b. In most cases the shape, size, and strength of the offspring is determined by the mother body, due to its longer influence on the new-born, more influence than the father body.

   The offspring of the first mating of a Swiss bull with a cow of our country breed—a Negretti male sheep rich in skin folds with a mother sheep of the country breed—is as a rule certainly more reminiscent of the maternal than of the paternal sire in terms of form, size, and body structure.

   Several empirical remarks of (now deceased) Prof. Sturm makes it highly likely that the father has a greater share in certain characteristics of the offspring, while the mother's share is more significant for others. Besides the clearer appearances of the bastards, the sire takes a predominant influence on the formation of the offspring front part, because, in addition to the head form, the mule and the hinny also have the well-known voice in common with their sires. This heredity is also confirmed by the fact that thick-headed calves are difficult to give birth to by our land cows if the latter are mated carelessly with bulls of very large beat. However, due to lack of numerous, often repeated and correct observations, we are by far not yet capable of a decisive judgment on this heredity, and can, therefore, at most recommend the hints given here for closer examination by contemporary and future generations.

c. In some cases, the reproductive capacity also suffers noticeable changes.

The English long-wavy rams, placed in Moravia, Fulnek in the beautiful Mr. Baron v. Badenfeld sheepfolds, continuously failed to mate with the rutting merino mothers.

It was Büffon who already remarked that there were far more female than male births in a cattle breeding in decline.

Likewise, in poorly managed livestock farming, it is more common to notice unsuccessful, infertile mattings and frequent untimely births are more common for poorly managed cattle breeding.

The travel reports of a Spaniard to his government indicate that in 1791, 1033 mares had no more than 73 mares and 2 colts out in the first area; in a second area, 654 mares—only 58 mares; in the third area with 3691 and in the fourth one with 1278 mares—only a few mares and no colts.

In my youthful years I have observed a similar, but less significant discrepancy between the number of mating and mature births, between mares and colts in the common horse breeding of our fatherland.

In contrast, the famous Mezőhegyes stud farm, founded by the Emperor Joseph in 1785, had 717 mares in Hungary in 1805:

	Foals	
	Colts	Mares
2 years old	190	166
1 year old	169	172
Lactating	162	154

One also often notices an unusually frequent retention or rejection of the sheep's milk in poorly managed sheepfolds.

d. In contrast to the general rule, one can sometimes see that the born animal is strikingly similar to the one of the mated genera only for causes that can be at most assumed but not proved.

§6

It is generally believed that the Lord of Nature made only two primordial animals of different genera, which were, however, similar in the main characteristics as to ensure the propagation of each animal species over the Earth. In these original parents of every animal species, the hereditary ability was already present to keep the essential characteristics of the species in the offspring, according to special circumstances, but also to take on new, more accidental characteristics.

The two genera of the horse, cattle, sheep, etc. will always produce a young one of the same species with the inherited essential characteristics by fertile mating. However, in their geographical distribution over the Earth, horse, cattle, sheep, etc. have been so varied in their accidental characteristics that the original form can no longer be detected in any of them one, and even more so to be proven.

§ 7

The circumstances that have caused, and keep causing, the deviations of the offspring in the characteristics from the ones of the progenitors, can be found partly outside the progenitors in external influences, partly in the progenitors themselves.

§ 8

The external causes for deviations may comprise the following:

1. The general climate, the region or zone.

   The sheep living in the hot zone has simple fine hair; the sheep of the temperate zone has wool; the sheep of the colder zone has a simply curled basic wool and a stiff, simple overcoat protruding from it. Besides, shape, color, horn, tail, and abundance of hair are highly modified. Horses, cattle, pigs, etc. show similar variations according to the zones too.

2. The particular or local climate, depending on whether it is more or less favorable to the nature of the organic body, more or less conducive to its flourishing in any particular respect.

   The sheep living in dry regions will never attain the size and fertility either retains in its offspring, which is what we admire in the sheep of the wet marshland.

3. The location in the natural state, depending on whether this grants the organic body the conditions of uniform

nutrition and protection against the weather change in a prosperous measure.

Following their various dispersals over the Earth, animals in the wild often experience indescribable hardship in their chosen habitats that last weeks and months due to the change of weather and seasons. In the period of growth, this hardship cannot remain without consequences for the animal's own physique, and during the mother's period of gestation, it cannot remain without consequences for the young in the womb.

Three aforementioned external causes of deviation from the original form enable us to explain why in the nomadic or blowing economy of humankind, the semi-wild animals almost always retain a uniform size, color, and form, because they all develop and grow under the same influences of the same climate and location. You can see this if you look at the horses from the semi-wild stud farms in Bukovina, Moldavia, and Wallachia, the Polish and Hungarian cattle herded to us, the Hungarian racket sheep, on the Syrmian pigs, etc.

4. Human care, in so far as it supports the nature of organic bodies in the formation process, gives their effects another more advantageous direction, or hinders them clumsily.

Man proves his influence on the offspring of organic bodies as a product of procreation.

a. by selecting the genera for mating according to their characteristics;

b. by altering the natural time of fertilization;

c. by the protection or abuse of the procreative power; by the enhancement or weakening of the procreative instinct;

d. by taking care to ensure that the nourishment he gives to the procreators and to the procreated;

e. by means of protection against adverse external influences;

f. by his cure of pathological conditions.

§ 9

The deviation causes that lie in the progenitors themselves may include:

1. The age of the genera the procreated descends from. The onset and duration of procreative capacity is often very different in the genera and species of organic bodies, and even in different individuals of the same species. In most cases, procreative capacity occurs naturally as soon as the organic body has completed at least three-fourths of its individually possible growth; a moderate satisfaction of the procreative instinct with sufficient care and nourishment does not hinder the completion of growth. It therefore occurs earlier in organic bodies with smaller growth terminated sooner, and later for the ones with large and later terminated growth. Examples in the plant kingdom comprise annual plants as compared to biennials, shrubs vs. trees, dwarf trees vs. tall-trunked fruit trees. In the animal kingdom, the comparison is made among hare, dog, deer, and stag. As for domestic breeding, the livestock of pigs, sheep, cattle, and horses crippled as a result of poor care are compared with those that grow longer and also bigger when well cared for.

   For those who takes offense at the rule on the occurrence of procreation, for example, because it contradicts the principles of other scholars, we would like to refer not only to the authority of the highly deserving professor Burger (see his Textbook on Agricultural Science, part 2, page 190), but especially to the usual state of nature in the wild, to the usual state in one's own fatherland.

   In the free state of nature, the genera mate as soon as the procreation instinct becomes strongly and vigorously active during natural growth; that is the case for bodies with a short growing period after the growth is completed, and for others with a long growing period before completion of growth. Examples comprise the birds, the roe, and the deer. In the latter, the narrow goat, the hind calf, the pricket make use of the procreative instinct, which has become active even before growth is complete, without any disadvantageous consequences for further development.

In the regions of Lower Austria, Upper Austria, Styria, Tyrol, and Salzburg known for excellent cattle breeding, no reasonable farmer takes decency to use the strong and perfectly grown-up young bull of 1 1/2, 1 1/4 and even barely up to three years to a moderate number of jumps; the female calf can mate at the age of 1 1/2, when the mating instinct becomes strong and repeatedly active. With the regular and careful care that is customary there throughout the whole life of the animals, there is not the slightest trace of any of the disadvantages that we would have to worry about in our care of the native bark from such an early mating season. Given the regular and diligent care during the whole life of the animals, which is common there, there is no slightest trace of any disadvantages that we would have to worry about in our care of the homeland bark from such an early mating season. There are excellent, beautifully built cows, small and young slender bulls there, because they are not allowed growing large as such; and there are also oxen of a size and strength, such as the one still looks for in vain here in the semi-wild, blowing economy.

We use 3-year-old bulls, 2 1/2-year-old calves, 2-year-old rams, 2-year-old sheep, etc. for the offspring. I wonder if these animals are fully grown already at this age. Does a 5- to 6-year-old horse, cattle, sheep, etc. fully complete its growth? If, despite this customary later use of the procreation instinct, our domestic animals become crippled, shrunk, and reduced in growth as compared to the ones of other regions, the reason is to be sought not in mere use, but in the procreation instinct abuse and in the bad, often very uneven care.

For organic bodies, the duration of the procreative capacity depends on their health condition, vigorous organization, the use of the procreative power, and on individuality.

The sickly or diseased animal has to struggle with causes of disease, overrides the procreation instinct, or expresses it in a weaker degree, suffers when using it, and should be excluded from using of the procreation instinct due to its own condition and the offspring to whom predispositions to illness are passed on.

Strong organisms retain the ability to procreate longer than naturally weak organisms.

Yet, the procreative power is exhausted earlier in case of improper use, which is true even for strong organisms. If in practice we sometimes see 100 or more mothers assigned to the strongest ram, and 100 or more cows in a short period of rutting to the bulls, as happens in the congregational herds, we must no longer wonder why, following such an improper use of a ram or bull, the mothers either remain unfertilized later on, or give birth to a weak young one with a tendency to malformation or disease.

Some animals of special individuality really do have an excellent procreative power, both in terms of strength and duration, and often achieve unusual result without suffering from it. If its numerous offspring are at the same time suitable for agricultural purposes, such a male animal is good for numerous female herds and is a true treasure for the farmer. In his textbook and the Möglin annals Sturm lists achievements of individual rams that are nearly unbelievable. I saw a 20-year-old stallion in the neighborhood just a few years ago that had been kept as a prancing animal for so long only because of its fertile procreative capacity. As we have already mentioned, what individual bulls have to achieve in our often very large village livestock, is admittedly not to the advantage of the offspring.

It is the farmer's business to study the nature of his domestic animals in this respect too, so as not to use the procreative instinct too soon, too long, or too often.

Some points mentioned here may be applicable to flora too.

2. The vitality, lifeforce, fullness of life, what others also call energy, which the procreative bodies may have even after time in very unequal degrees according to age, health, voluptuousness.

The greater this natural fullness of life, the more vigorous and numerous the offspring will be; the less it is, the procreation will more often be infertile, the offspring more fragile.

To prevent the disadvantages of the improper use of the procreative instinct in irrational animals, nature restricts the life-state excitement for the purpose of procreation to certain periods, which we, therefore, call the rutting or mating season; for plants it is the time of flowering.

People use this natural time of procreation when it suits their purposes or change it with altered care when it does not. They increase the vital activity of the genera in this time with better care and the preceding long separation of the genera.

People are careful not to reduce the vital activity by abuse of the procreative instinct or applied coercion.

3. The dissimilarity in the characteristics of the two genera mated together, each of them passes a part of its characteristics on to the offspring. The offspring must, therefore, differ in its characteristics as compared to each of its progenitors.

Excessive dissimilarity in characteristics is often one of the main reasons why the genera fail to mate, why the accomplished mating is often unsuccessful, or why the actual offspring is often more misshapen than any of its parents considered separately.

4. Sometimes there is a great similarity between the genera, namely, when both of them have a common defect, which is contrary to the economic purposes and can be inherited, or also an advantage favorable to the economic purposes, and that now, of course, appears far more strongly and distinctly in the offspring of such a procreation than it used to be previously perceptible in either of the parents.

If a ram and a mother sheep, mated together, were poor in wool due to a defective function of that skin layer the hair grows from, their offspring would be even poorer in wool than any of its progenitors.

We do not know all the defects or even the good characteristics passed on from the parents to the offspring yet, nor do we know them precisely enough due to lack of correct and often repeated experience. In general, we only know that defects in the organization, weakness and deficient activity of individual vascular and the entire vascular systems, but also advantageous characteristics of the same, are passed on to the offspring more easily if they are common to both genera used for mating.

5. Previous fertilization of the mother bodies by individuals of the opposite sex, but very significant dissimilarity in characteristics.

If a horse mare has been mated one or more times with a donkey stallion to give birth to the mules, it remains in her disposition to develop a more or less donkey-like fill, even if the horse mare is later actually fertilized by a horse colt. Even in the plant kingdom, for example, one reason for the variety of table potatoes is said to be found in the fact that table potatoes and cattle potatoes often stand side by side in the field and mutually fertilize each other by sharing the pollen. If it were also proved as really probable, it would be the proof that fertilization has a lasting influence not only on the fruit but also on the mother's body.

These crossbreeding phenomena in the animal and plant kingdom indicate similar successes when genera of the same kind, but of very dissimilar characteristics, have been previously mated with each other, and subsequently the procreation product of two genera of the same kind nevertheless reverts to the former dissimilarity.

§ 10

The offspring is known as a variety of two genera of the same kind and similarity in characteristics beget an offspring that differs from its producers in the less essential characteristics, for reasons which lie either in the producers or in external circumstances.

For example, a hornless cow is an offspring of a horned bull and a horned cow; therefore, it is a variety. Thus, we get new varieties from the potato seed and other plants due to causes that are often unknown.

In the plant kingdom, the preservation of an appreciable variety, i.e., its persistence in the offspring is quite easy if propagation by growth (direct or indirect) applies. As for the animal kingdom, the preservation of a variety is more difficult and depends on multiple circumstances.

§ 11

If the deviation cause of the offspring from the characteristics of the producers is permanent and general, not only will the variety be repeated in the offspring of like producers, but the deviating characteristics will, with the persistence of the cause of change, become fixed and will be stronger in the new offspring of the same.

Cattle may be a good example, because they are driven from the Tyrolean mountain regions with their special characteristics into our fatherland and our stables. They face entirely different climatic conditions here, an unaccustomedly worse, often very poor care. In these changed circumstances, the breeding animals that are brought in suffer and give birth to calves that are not as strong as compared to the ones that would have been born in their original homeland. Since the causes, why the driven-in stock animals are stunted, have influence on all calves born here too, the latter will be stunted like their ancestors, or even more. We therefore obtain a deviating variety, whose again mated genera, with the continuance of the deviation cause, give birth to young ones deviating even more from the characteristics of the progenitors, until finally—with continued offspring mating in the offspring line and due to continuance of the deviation causes, every trace of the peculiar characteristics of the progenitors disappears forming a cattle strain with characteristics suitable to the conditions of the climate and negligent care. This is the way how almost every trace of a Swiss cattle breed with blackish tiger-like markings on a light ground has disappeared, as it was brought to our states about 100 years ago. The remnant of this breed still shows its origin at most by the coloring and the short, fine, sparse hair, but the shape itself has degenerated into the usual country breed.

§ 12

If, during the continuance of the deviation cause, by long-continued mating of similar genera, offspring are finally produced from an original variety, which nevertheless preserved the deviating, enhanced characteristics of the variety and that can still be exterminated, even if the cause for deviation has long ceased ever since, a subspecies with hereditary characteristics has been formed from a mere original variety.

In this way, we have at the same time already indicated the way how a mere original variety can be elevated to a subspecies with hereditary characteristics.

An accidentally born hornless cow is a variety we do not know the deviation cause for. If we have two hornless genera of cattle, and always mate them together, several, although not all, offspring will be born. Now again two hornless genera from this first mating are continued in the second, third, fourth, and so on. The descent will continue and, with the increase of descents, the original new characteristic will become more and more fixed in the offspring; and they will finally become so solid that, even if in the later descents, hornless cows are mated with horned ones again, hornless cows will, however, still appear in the herd for a long time to come.

§ 13

For domesticated animals, we call the subspecies common breeds, for seedlings from the plant kingdom we also call the subspecies varieties.

We understand breeds to mean a number of domesticated animals of the same type with certain pronounced, assuredly inherited characteristics which are more or less missing in all other animals of the same type.

Those who are motivated by our restricting the use of the term breed just to animals that are cultivated under the influence of humans and therefore referred to as domesticated animals would want to give us a few examples of how the term breed can be applied to wild (not feral) animals.

However, the terms of variations and breeds complement each other. Whereas the heritability of deviation certainly occurs in the continuation of the cause of changes in the variation since the breed starts. Whereas the heritability of the deviation fluctuates or ends entirely with the cessation of the cause of changes, new causes of change trigger new deviations, since the term variation starts.

Examples of breeds can be seen in the white and black human races, the light Arabian, the cumbersome originally German horse breeds, the Polish, Hungarian breeds, the Tyrolean, Swiss mountain breeds of cattle, the racka [sheep], the fat-tailed merino and marsh sheep, the English long-wooled sheep, etc.

§ 14

When the genders of one species but different breeds are mated together, then the offspring are not entirely similar to its progenitors, rather only partially similar to the paternal progenitor and partially similar to the maternal progenitor.

The offspring of two genders of one species but belonging to different breeds is called a mongrel, also metis, mestize, sometimes also bastard in the figurative sense.

The mating between the genders of two breeds belonging to the same species is called a crossbreed.

The deliberate combination of genders of dissimilar breeds for the purpose of mating is called crossbreeding or crossing.

The combination of Tirolean steer with a rutting cow from native stock is a crossbreeding between both breeds; their mating a crossbreed and the progeny of this pairing a mongrel.

One usually takes shelter in numbers to express the degree of similarity between the mongrel and his disparate progenitors. One makes the particular characteristics of a gender equal to 100, and the deviating special characteristics of the other gender mated with it equal to 0 in comparison with the characteristics of the first. Because both genders have the same proportion based on the general assumption of the characteristics of the mongrel, the success of the crossbreeding in the first descent is expressed with the following equation:

$$\frac{100+0}{2} = \frac{100}{2} = 50$$

The quotient 50 indicates that the characteristics of the first mongrel are equal to half of that from the paternal progenitor and half from the maternal progenitor. However, we have already become aware how much the success of the procreation against all probability in some cases is altered by unknown causes. For this reason, the mathematical equation is not a means of verifying accuracy, rather only an aid in clarifying the probable success.

§ 15

The crossbreed is a welcome means of gradually replacing the characteristics of a given breed, either partially or entirely, and in their place gradually transferring and affixing the other characteristics for another breed of the same species with continued mating of the genders either partially or entirely. The beneficial modification of characteristics of a breed by crossbreeding with another breed of other characteristics is called grafting by crossbreeding.

On the way to crossing the native horses of England with the horses from Arabia and the Barbary Coast, the famous English racehorse was formed. In the same manner, the native landscape was refined with merino sheep and the native cattle with cattle from Switzerland, Tyrol, März Valley, etc.

§ 16

In order to achieve the purpose of crossbreeding, namely, to refine a breed in terms of characteristics to the most economic manner, the following principles must be considered in the application.

1. One selects the male animal from that breed for which the refinement is to begin and mates it with the female animal of that breed through which it should be refined.

   One obtains slightly fewer male animals from foreign breeds to mate with many of the available female animals than vice versa. Also, with slightly fewer male animals used for crossbreeding, one gets farther sooner and more easily to a homogenous herd than vice versa. The refinement from female animals from the foreign breed would certainly be faster but more expensive to achieve the goals.

2. One ensures, as much as is possible, that the male animals to be chosen are truly breed animals and not mongrels obtained through crossbreeding.

   In horse breeding, one is on this point sometimes so anxious that one presents bloodlines for famous horses that reach back 2000 years. Also in England, one goes back 300–400 years to pedigrees from famous horses and let the jumping and foaling confirm famous mares with sires and documents.

When purchasing the breeding ram, you may well exercise caution. If you believe the assurances of the seller, they have any number of breed animals. Why do these men not present their sheep farm registry to the buyer for verification? Why does the owner of well-known herds not put the history of said herd in writing for general verification by others in public gazettes?

3. In mongrels, the existing characteristics are not yet so fixed that the certain heritability of the same in the descendants is left to hope.

One considers whether breeds also are truly suited to beneficial crossing, and whether or not their descents will come out worse than the animals of that breed that one wants to refine with crossbreeding.

The cumbersome colossal carriage horse, as he threatens to disappear soon from what remains in Styria and Salzburg, has long been a highly valuable breed animal, as long as we don't have sufficient waterways or railroads. Would it not be utter nonsense to want to ruin this horse with a lean, light-footed, Arabian stallion?

Likewise, it would ruin the sheep breeds if you wanted to crossbreed our short and fine-wooled merino sheep with the long-wool English, or the small Heidschnucke with the Marsh sheep.

4. The sheep breeder must himself be clear about which characteristics and to what degree he wants them to be transferred to his breeding animals in order to then select and mate the female as well as the male animals according to the grafting purposes intended.

He who does not clearly know what he wants, will blunder about with uncertainty in the selection of resources which will be evidenced in numerous errors of judgment in the grafting of our domesticated animals, especially in the breeding of horses and sheep.

5. The more the mothers which are to be grafted are similar to the males to be used for the crossbreeding, the likely and easily the goal of the grafting will be reached. And

vice versa, the less they are similar, the more difficult and later the goal will be reached.

It was an essential advantage for the improvement of our domestic sheep breeding that our native sheep were somewhat grafted before the introduction of the merino sheep by Paduan sheep and therefore were better prepared for crossbreeding with the merinos.

It was a harmful error of judgment to use the merino ram in several sheep farms in Hungary to crossbreed with the Racka.

The path to the goals of the determined grafting is very different in the length, depending on whether the grafting of the mother starts from the Racka, the native sheep, or from a somewhat grafted breed.

6. With vigilance and attentiveness, as at the start of the grafting, their progress must be observed and the means for their promotion handled tenaciously.

To ensure and promote the grafting, refreshing, or the new long enough continued crossbreeding with new animals of the same breed, starts from the grafting.

In many of our current sheep farms, the grafting of the native sheep with merino rams was started almost at the same time in the last three decades of the previous century, but with what uncertain and uneven success! One can usually mislead the first noticeable progress in grafting, remain at the halfway point, and then go backward again in the grafting as quickly as one came forward.

Just how easy such a deception is possible, and is excusable in beginners in sheep breeding, can be explained again with the mathematical equation above, when we carry it out for several descents.

Degree of descent

$$\text{I. } \frac{100+0}{2} = \frac{100}{2} = 50$$

$$\text{II. } \frac{100+50}{2} = \frac{150}{2} = 75$$

$$\text{III. } \frac{100+75}{2} = \frac{175}{2} = 87\frac{1}{2}$$

$$\text{IV. } \frac{100+87\frac{1}{2}}{2} = \frac{187\frac{1}{2}}{2} = 93\frac{3}{4}$$

$$\text{V. } \frac{100+93\frac{3}{4}}{2} = \frac{193\frac{3}{4}}{2} = 96\frac{7}{8}$$

$$\text{VI. } \frac{100+96\frac{7}{8}}{2} = \frac{196\frac{7}{8}}{2} = 98\frac{7}{18}$$

Here the quotients 50, 75, 87½, 93¾, 96⅞, 98⁷⁄₁₆ the probable progress of grafting with crossbreeding. In the early days of our grafted sheep breeding, one could, and still today can be deceived with the dazzling success of the third descent even more so because you could believe the apparently great approximation of completing the grafting will save from additional costly ram purchases. Even faculties were taught the principle by very learned teachers in 1807: "For native female sheep, already of better wool, the grafting with crossbreeding with merino rams in the third generation (descent) is complete, for coarse wool native sheep however, the completion of the grafting will occur only in the fourth or fifth generation." Still today, we find in the 3rd Edition of our public presentations guidelines presented from Trautmann, 1825 in § 1855, the incorrect tenet unchanged: "For the most cost-effective circumstances, as they rarely occur, the continued crossbreeding of the native sheep with pure merino rams in the sixth descent suppress the characteristics of the native breed and transfers those of the merino breed however not yet so fixed so profoundly in the organism of the mongrel that one now in the use of their own mongrel ram has to be concerned about not abundant relapses to the characteristics of the original native breed, even the gradual setback of the entire herd.

When the continued crossbreeding transfers and fixes the characteristics of the grafting breed to the progeny of the grafted breed that only a new refreshment is needed from time to time in order to keep the characteristics hereditary in additional progeny and to avoid setbacks, then the current descent is called by some half stock [*illegitimate*].

The name half stock seems to us not very correctly selected for such grafted animals because the word indicates, based on its composition, that the descents are similar to half of the male and half of the female progenitors and therefore have reached a high degree of suitability for heredity. However, we have not found a better expression and therefore need to keep the suggestion until a better one is discovered. Petri in his work "All About Sheep Breeding", 2nd Edition 1826, explained that in grafted sheep only some animals are half stock, originating from crossbreeding continued through 16 generations with pure breed animals. If you estimate almost three years for one generation, then we need more than 40 years to graft a native stock of female sheep up to half stock.

In this somewhat strict sense of half stock, the sheep breeders who frequently offer breed animals for sale have shown few half stock sheep.

Strictly speaking, perhaps the word stock which indicates a certain hereditary quality in the construction of bones, muscles, and tendons, may not even be applicable in sheep breeding because this is more about the construction of the wool hair which can definitely be changed more easily than the entire structure of the horse.

If the characteristics of the grafting breed are in this way merged with and fixed in the grafted breed through continued crossbreeding that the genders of the same without new refreshment from the grafting breed in mating with each other transfer their characteristics to the descents with complete certainty, then the half stock is reshaped into a hereditary stock.

The hereditary stock is then only named breed when its animals carry certain distinctive valued characteristics.

Petri postulates crossbreeding 25–30 generations to form a hereditary stock, so more than 150 years for horses and almost 70 years for sheep. The strength of this claim is confirmed with the history of horse breeding, and in sheep breeding with everyday experience that our sheep herds soon degenerate again as soon as we take away very prudent management and hand over the everyday handling to the common sheep handler.

1. Male descendants from the grafted herd with noticeable setbacks to ungrafted breeds are intentionally separated from the herd sooner or later and not used for further breeding.

   Female descendants from the grafter herd with noticeable setbacks to the ungrafted herd are intentionally separated out from the herd sooner or later and not used for further offspring.

2. Every mating with other male animals that are not purely descended from the breed from which the grafting started must be carefully avoided.

   The notably large inequality of animals in so many expensive stallions is primarily based in the frequent transgressions against them and the previous principle.

3. The purpose of the grafting must be supported by well-considered cultivation which the farmer has provided to his seedlings in every period of lie from creation to separation from breeding.

§ 17

Because several purchases of a large number of male animals for the purpose of grafting numerous herds is costly, one prefers to buy a proportionate number of male and female animals to obtain the necessary male animals of that breed with which one wants to crossbreed his own female animals.

A proportionate number of male and female breed animals are called a main herd.

This main herd is bred just among itself without mixing in other breed animals, grows larger over the years from the offspring of female animals, and supplies the necessary male breeding animals to crossbreed with indigenous.

If we assume the practical case, someone has agreed to grafting 500 sheep with merino rams which he will continue through 10 generations (for which he needs at least 25 years for the greatest economy with the use of the animals) without mixing in foreign impure blood. When he casually allocates 30 females to every ram for breeding, and it remains serviceable for six years as a breeding animal, at the beginning the grafting of 16 rams and through 25 years of the allocated grafting time 64 rams that may cost 10000 Fl. C. M. per current prices for better breeding rams.

If at the start of the grafting, he buys a main herd of 50 females, this should cost 1000 Fl. C. coins, furthermore, to graft the 500 female sheep and to cover the 50 main females with 16 rams at 2500 Fl. Using the information above he saves 6500 Fl. C. M. and at the end of 25 years will have a very considerable significant main herd of pure breed cattle.

§ 18

The continued cultivation of a main herd by mating the genders with each other without mixing in outside blood, is called inbreeding, and this way of continued cultivation itself inbreeding or also kinship breeding because it is unavoidable in a moderate main herd the animals will not mate together which are in the same relationship from their descents, like brother and sister, parent and child, cousins, aunts and male cousins male family lives.

§ 19

In earlier times in all of Germany, mating the related genders of a breed was considered more disadvantageous the more closely related they were. This opinion was held the longest in the Austrian states where they are still defended today in the exhibited textbook by Trautmann, 3rd Edition 1823, p. 231. The following reasons are given:

1. The mating of relations is not according to nature, but instead contrary to nature because natural freedom the animals always choose outsiders of their species for mating.

   Hengste often bristles against mating with the mothers, bitches in heat against mating with brothers and sons, cows and sheep prefer to choose foreign bulls and rams for mating. Young deer bitches, fawns, birds from the same nest remain together until mating time but then they separate. For plants, this switch of mating is operated by insects and wind.

2. If blood relatives really mate with each other, then they create weaklings that already carry the seed for degeneration in them, oftentimes remaining barren or not able to carry embryos to maturity.

In stud farms, the entire horse breed degenerates when there is not a switch to stallions from 4 to 4 years. A switch with the bulls from herd to herd has always resulted in more and more and prettier calves for the Hungarian Pusten.

For the Englishman Prinsep, large cows became smaller in the descendants by mating with related animals. The Englishman Sebright attempted with dogs, doves, and other birds to confirm this degeneration. Pigs cultivated for many years with blood relatives at an English tenant farm descended at the end entirely from instinctive procreation, or brought crippled young ones into the world. The attempts of the Englishman Knight with plants likewise confirm that the descents or unrelated genders are stronger and more durable than the related genders.

§ 20

However, there have at all times been practicing farmers that not only confirmed all disadvantages, but also even with obvious benefit in the breeding of domesticated animals, inbreeding or relationship breeding. Even in open nature, traces of relationship breeding can be seen.

We refer as proof to the facts from many years of agricultural practice:

a. In the mountain areas of Lower, Upper, and Inner Austria, where the economies are scattered across very isolated mountain slopes and the field work is done with oxen they have raised themselves, for centuries, the farmers have let their few cows be mounted by an almost two-year-old male descent of the same, because during the third year, he is castrated to become a plow ox. I myself lived for six years in such a mountain region on the border of Styria.

Among the mountain people, I have not only never heard a complaint about the resistance of the relate animals, about the *Geltebleiben* of the cows, and the degeneration of their progeny, but rather have to acknowledge the truth for an assessment that there the lovely type of März Valley horned cattle is at home, and maintains itself in progeny with the same perfection.

b. In the cattle stalls of the grand estates here, for many inconsiderable years no decency whatsoever has been taken to raise the well-built, growing, strong bull calf from the estate's own herd to breeding animal to mate with his female relatives and for longer serviceability with his descents.

c. In the cattle herds of the communities, this breeding with related animals is even more common.

d. In England, the famous landlord Bakewell and his successors mated cattle, sheep, pigs in very close relationship just for the purpose of obtaining animals of a certain shape, size, capacity for fattening, tenderness in bone structure.

e. In Spain, the rams from one's own herd have always been used in the roaming herds for mating without fearing any kind of disadvantage due the relationship.

f. In the most famous Saxon sheep herds, in the most famous herds of our own fatherland, despite the warnings so frequently originating from Vienna against the disadvantages of running relationship breeding for so long that you cannot find any better ram than from your own herd that you select him anyway because he has better characteristics and can be transferred to the herd.

g. For pig and bird breeding, for years, the related genders have been mated without all the disadvantages in all economies.

h. Government Advisor Mr. Jordan, probably through the authority of Professor Pessina from Czechorad, emphasized, at least taught at the university in Vienna that relationship breeding has disadvantages, but contradicted himself with his suggestion for an experimental farm in Bösendorf in 1810 with a main herd of Swiss horned cattle for cultivation with itself, and was maintained until the dissolution of the experimental farm in 1824.

i. Trautmann himself (as ostensible author of our hand-book) speaks on page 324 II. B. of cultivation of the native cattle stock in and with itself as a feasible and performed method for grafting the same.

Further, we refer as proof to an entirely harmless breeding with relatives actually in the wild:

a. It is known that many migratory birds, like swallows, quails, larks, storks, nightingales, etc., every time in their return in springtime, like to select the same region for the summer stay in which they originated. Might this persistent purpose not frequently result in mating of relatives, especially in those birds that are often in very small numbers in our home region, like storks, nightingales, cuckoos, etc.?

b. It is known that the non-migratory birds of one and the same region, like goldfinches, siskins, finches, linnets, etc., live in companionship after mating season, that leaving the nest in the fall for a mutual group that gather to go through the winter together in the same region. It is known that deer from the same region as the so-called sedentary game mostly live together in packs and staunchly hold to the region where they were born if they are not driven from it out of fear of humans or unsatisfactory sex drive. Should the later separation of the group into pairs or smaller groups not result in mating with relatives whatsoever?

Why do the few lines of deer with which a zoo is beset multiply without degeneration, few pairs of doves with which a dovecote is occupied in immense number if the external conditions would assure their flourishing?

c. Do no matings of relatives occur in animals whose harmful propagation has been limited due to their rapacity, the wisdom of nature, or by man and they therefore occur in very low numbers, like bears, wolves, foxes, badgers, weasels, eagles, vultures, etc.?

d. Do we not derive the existing of all animals from two original parents and thank their origin in the acclimatization by man countless descents in the entire world not a few original animals multiplied with relatives?

e. In the plant kingdom, breeding of relatives occurs so frequently that most plants fertilize themselves as hermaphrodites and that the plants standing next to each other in the natural condition with entirely different genders sprouted from the seed of one and the same mother plant.

When in some studs the entire horse breed degenerates, the cause will be found in that animals have been foolishly selected for mating that do not go together.

§ 21

Both contradictory opinions, that breeding with relatives is unnatural and harmful—"breeding with relatives could even be useful and by far not as unnatural as one thinks" —can be unified when we verify the conditions under which the mating between closely related genders can be harmful or unharmful. This includes:

1. Consider the physical structure of the genders that should go together.

   If both genders are flawless, the descents then will with greater probability also be flawless, with consideration of whether they are related to each other or not.

   If both genders have a mutual flaw that does not hinder the economic purposes, it is then that much more likely to be passed on to the progeny from mating with relatives the longer the error is inherited in the ancestors.

   Therefore, in the breeding of domesticated animals, the selection and composition of related genders for mating is entrusted neither to their natural drive nor unskilled hands, but instead is a matter for the thinking cattle breeder.

2. The caution can exist, whether in the characteristics which are sought by cattle breeders to be increased in the offspring through intentional mating of related genders, the health of the animal, their other value, and the uninhibited completion of the natural daily life, or whether one is involved in apparent contradictions of nature.

Thus, a certain delicacy in bone structure, which is very often closely associated with speed for horses, for cows with milk productivity, for sheep with fine wool, for pigs with fattening ability, has its limits over which the delicacy degenerates into weakness and susceptibility to disease. It is a matter for the breeder not to exceed these outer limits.

In this way, a high-milk productivity rarely contributes to great strength for work or with great fattening ability; an abundance of wool not with excellent wool fineness; great fertility not with an unnaturally large physical structure, etc.

He who wants to breed related animals with a single pronounced characteristic must either avoid the contrasting characteristic or moderate the appearance of the predominant characteristic.

By working toward a one-sided goal in the breeding of relatives, the body of the descents can be reformed such that it either is unable or less able for procreation, for conception, for carrying, or for complete formation of offspring. The above-mentioned cows of the Englishman Prinsep, the pigs at the tenant farm were certainly animals mated for such work toward a one-sided goal.

In 1822, I saw a two-year-old pig from Austria; in fattened condition, it weighed over 800 pounds, and could stand on its weakened feet only with great effort; in 1824, I saw a not yet three-year-old bull born in Mähren of Tyrolean breed on which the navel region was barely 6 inches above ground and therefore was entirely unsuited for mating.

Such one-sided corruption are obvious contradictions against nature, which are induced by controlled mating of relatives resulting from capriciousness but can be avoided.

3. The influence of man on the procreative drive in his animals through weakness or increase of the same.

The procreative drive is weakened by the genders cohabitating and living together in the same stall space, in the same atmosphere filled with their own vapors, with poor care and nutrition.

The procreative drive is increased by separating the genders and removing them the same from the atmosphere filled with mutual vapors.

As a result, the above-mentioned experience cases explain how one can seek to prove the harm and the unnaturalness of breeding of relatives.

§ 22

Among the associated precautions, breeding of relatives remains an unreplaceable means of reshaping the bodies of our domesticated animals based on discretion, to change their characteristics based on the various economic intentions, or to keep them unchanged in the progeny.

The damage is immense and incalculable as the prejudice against breeding with relatives has fomented so long and extensively in our sheep breeding. With the same, most of our sheep breeders constantly fall into the hands of the shouters and agents with breeding rams. One believes, and many still believe today, that everything has been done if you just avoid breeding with relatives with the purchase of new rams from this and that herd.

§ 23

Pure breeding is separate from relatives, which consists of always mating animals from the same breed without consideration of blood relation.

He who can have a better breeding animal in a foreign herd from the same breed at acceptable prices then he has himself, would be a fool if he wanted to scorn breeding of relatives out of stubbornness.

He who has corrupted the characteristics of the offspring in breeding of relatives with incorrect combination of genders, will see the resulting errors the most quickly when he chooses flawless breeding animals of the same breed from a foreign herd, meaning practices pure breeding.

Appendix 6—F. Diebl's Lecture About the Formation of Wheat Varieties

1829. *Announcements of the Imperial-Royal Moravian-Silesian Society for the Promotion of Agriculture, Natural and Regional Studies in Brno* 23:177–179

Notes on the views expressed by Baron von Witten regarding various types of wheat. (№ 40 of the release dated 1828)

In his work addressed the horticultural society, Mr. Baron von Witten has made remarks against the agriculturalists where he accuses them, on the one hand, of an excessive self-righteousness and, on the other hand, of a gross ignorance with regard to different varieties of wheat, **the origin of which lies in one and the same variety of wheat and ascribes to them erroneous doctrines that no single professor of agriculture in the entire Austrian Empire admits,** while at the same time he himself does not display the clearest insights appropriate to the state of science.

Had the esteemed Baron von Witten taken the small trouble to consult the textbooks and journals **on agricultural theory written in these states by professors and other scholars,** e.g., by the [*illegible*] Leopold Trautmann, Subernialrathe Burger, even the one of prof. Sturm and the [*illegible*] prof. Fohl's archive, and many more, he would have been convinced of the opposite, and would have probably gained some knowledge of his own on this subject. I am a temporary occupant of an economic teaching post in this country and am thus called upon to present my views the presentation of my teachings on this very important subject of agriculture is based on. **According to the most recent authors, the natural history, which must be regarded as the basis of agricultural theory,** assumes six species of the genus *Triticum,* which cultivation for grain yield is used in agriculture, namely:

a. *Triticum commune,* common wheat

b. *T. turgidum,* puffy or English whitewash

c. *T. polonicum,* Gommer or Polish wheat

d. *T. compositum* Many-grain or miracle wheat, Maroffan wheat

e. *T. Spelta* v. Zea Spetz Defen

f. *Monococcon,* einkorn wheat

The natural historian chooses essential and permanent characteristics of the organizing bodies as the distinction between the species, the characteristics that pass on from the producers to the produced and that, if the same become imperishable under different climatic soil conditions or other local conditions, nevertheless re-emerge in the form appropriate to them. The cause of their development is still beyond the scope of our scientific knowledge, although physiology allows us to assume with a high degree of probability that they are the result of hybrid fertilization under the particular influence of the climate of the soil mixture.

Most of these species also have their forms, which distinguishing characteristics are no longer so permanent, but are subject to change as a result of changes in climate, soil mixture and other forces and substances influencing their vegetation and fertilization, while the varieties are also known as sorts, in so far as they reproduce permanently in the characteristic features of their form under the climatic and other conditions suitable to them.

Most probably, they owe their origin more to the influence of climate and soil mixture and other local circumstances than to unequal fertilization. The natural historian is concerned only with the permanent species in his classification and takes no further notice of the varieties. On the other hand, he makes the varieties a subject of fine scientific research by taking into account their prosperity and the higher yield or better quality resulting from it and strives to make himself a [*illegible*]. These varieties too can be assumed to be more the outgrowths of culture than of free-acting nature, which are produced anew in this way. Under the special influence of external circumstances that also influence fertilization, crops are sometimes formed deriving from those they originate from. The attentive and eager sand-grower draws these out and seeks to multiply them and, by doing so, forms a variety in the above-mentioned way, so that a **new variety arises. Some of them are more constant, while others are less constant and can be mixed with each other,** e.g., the white Kanna wheat, also called [*illegible*], the yellow land white, the red wheat that are most common here in the low mountain ranges.

The differences between the species lie mainly in the different formation of the grove of leaves and grains and in whether the glumes are attached to the grains or only enclose them loosely. The subspecies or sorts with different colors may comprise differently colored [*illegible*], glumes, awns and grains, as well the stamens and pistils many of which are sometimes bluish and reddish. The pistils are sometimes pale, ciliated or hairy, either green half-green or green pale. These species and sorts are classified as either winter, summer or alternative wheat, depending on their vegetation period or the time needed to complete development and reach maturity as soon as the climate of the temperate zones allows that, so it must only thrive with vegetation that is anti-ciliated in the preceding year before winter, and must then remain in the field over the winter, whence its name is derived.

Since, on the other hand, the second one has such a short vegetation period, the summer period is entirely sufficient for vegetation and maturity; the latter, however, is between the other two and, thus, requires a somewhat longer vegetation period than the second one and a bit [*illegible*] as compared to the first one. It is therefore just as suitable as winter crop and summer crop, the only difference being that it must be grown earlier as the former and [*illegible*] than the latter, but it is by no means limited to a single crop and can be grown several years in succession as a winter or summer crop; this is the basis for the difference between the earlier and the later ripening summer wheat. We have a collection of 46 species and varieties of wheat at the Cebranstalt of Biefig and, **during three years of practical use in agriculture, I have had the opportunity to observe how several of them have been transformed into others right before my eyes, i.e., within a short time, or have been degenerated. I am eager to** add the monograph of the species and varieties of wheat that I have found and my experience with them, as practical proof of the theory presented here, as soon as my regular occupation permits.

Franz Diebl
Suppl. Professor of Agricultural
Science and Natural History

Appendix 7—Hempel's Paper About Artificial Fertilization

1820. *Economic News and Announcements* **Nr. 21:163–164**

About the origin and the importance of the
different varieties of grain species

A lecture given in a private company by
enthusiasts of agriculture, from

Georg Carl Ludwig Hempel, pastor of Zedlitz

The knowledge about the changes and deviations of plants perceived in Nature by seeds of different varieties, but closely related in their genera, led naturalists to the artificial fertilization of the flowers in order to produce new varieties, and to guide the emergence of the varieties by means of art which one only accounted before as an accidental freak of nature merely from many dispersed seeds. Some florists use such artificial fertilization, namely with a brush the dust of the male's seed, which has been wisely selected to produce intended new shapes, is transferred on the female part of another flower, **thus new compositions of colors arise, through such production one is fertilizing and matching the plants, this is already often practiced with carnations, where one can even determine the outcomes in advance with fair degree of certainty:** after such combination a new carnation from the seeds of artificial insemination with a novel color must arise as we previously thought since we have chosen the male pollen from another flower with the intention of refinement desiring a new plant, where we have chosen the female of another different flower bearing in mind the creation of an ideally matching new plant. From the seed which has arisen from such refined **"artificial fertilization"**, a new progeny arises, composed of the properties of the father and mother flower. **In this formation it is important to know, which properties, e.g., shape, size, color is propagated from the male and**

female side each time to the newly generated progeny, and what
this child regularly appropriates from the father and mother
side in order to establish the intended development of the new
plant with certainty according such rules.

The scientific pomologists have already applied such artificial
fertilization to the fruit-tree blossoms for the production of new
varieties, and the English, and especially the noble Gentleman
Knight, the President of the London Horticultural Society, who
have made many remarkable experiments in this regard, and
from these experiments important results were obtained about
the artificial formation of new excellent types of fruit in the
future according to human ideas outlined in advance. According
to such information obtained in higher scientific pomology, we
have the greatest hope of uniting the most excellent properties
of two or more different varieties in one new plant produced by
artificial insemination, e.g., [we could] give the Borstorf apple the
size of the pound apple, bring the beautiful red color of the cherry
apple or to the dazzling whiteness of another apple together etc.
What a great triumph from higher scientific pomology would
be, to achieve that fruit types now developing under accidental
circumstances sometimes well, sometimes mediocre, sometimes poorly would now develop according to shape, size, taste,
and color, as one has previously ideally imagined it by using
such intellectual research and exploration such as those opened
to us by artificial insemination. The question now arises whether
artificial fertilization with different types of pollen of a plant species for the production of completely new, more advantageous
varieties can also be carried out for great practice in the case of
cereal species. To make the experiments with this intention and to
investigate the important new results to be expected in the case of
artificial fertilization of the cereal species, of course, a man with
deep botanical knowledge, of keen observational spirit and of
a tireless persistent patience is needed to grasp precisely the
intellect in these experiments and to hold firmly to it for further
clarification, since we cannot expect important results soon and
so easily from such critical experiments using artificial fertilization of cereals with different kinds of pollen. It would therefore
not be expedient choosing the production of new, more advantageous types of grains through soil and climate, I would rather
choose applying artificial fertilization to produce new types of

grain verities, which are more productive and useful, which is the subject of our discussion today. **Practical farmers, who are born naturalist by profession, cannot concern themselves with setting up and observing these laborious experiments.** And in general, the artificial fertilization of the flowers for the production of new, advantageous, productive types of grain with great advantages is currently still too high above our agricultural horizon as this new promising star is only hopefully shimmering to us from the blue distance.

It is enough beforehand if, through what has now been said, we can take a brighter look into the future's immeasurable realm for the benefit of fields, vegetables, fruits, and silviculture, through the artificial fertilization of the flowers to produce new and profitable varieties, which is revealed to us by this important experiment brought up for discussion, **from which this new abundant and ever-expanding source the greatest treasures will only emerge if one observes with sharpness through the brighter light of the increasing enlightenment and through often repeated observations attempting to penetrate deeper into the matter.**

It is more expedient that the subject to be discussed jointly is taken from the area of what is already known and generally understandable, but which should be in precise connection with what has been said above. I am therefore limiting myself only to the different types of grains, which actually already exist in advantageous cultivation; they may have received their peculiarities and differences from the soil, from the climate, or from a fertilization obtained with the pollen of a somewhat different seed.

The variety of sorts is of no insignificant importance in the cultivation of grain. In fruit-growing, a precise knowledge of the varieties has such an important influence on the prosperity and yield that anyone who plants an orchard can be assured in advance, at least in our northern regions, of twice the profit what he was expecting, if he selects the varieties according to precise pomological expertise. Knowledge of their peculiar portability and local prosperity as if one merely accidentally took them for cultivation without proper knowledge. In the case of the various sorts of cereals and fodder herbs, the importance and the advantage of their knowledge and selection for practical, advantageous cultivation do not strike us as clearly as in the case of fruits, but here the varieties are not yet so generally known and not tested properly for their true benefits. **I also noticed from my experience that one of my friends doubled the annual yield of *peas* by choosing an early English variety that was rich in abundance, and that they flourished very well by blooming before the thunderstorms occurred. [...]**

Notes

1. The word what Ehrenfels is using here is *"genetische Kraft"*, this is a direct reference to German *Naturphilosophie*. The Author.
2. Originally wool processing started with fulling, which was carried out by pounding of the woolen fabric with a club, or with the fuller's feel or hands. Urine, as a source of ammonium salts, was used for cleansing and whitening the cloth. The Author.
3. French term for runaways. The Author.
4. Hungarian term for groom or stableman. The Author.
5. Hungarian currency. The Author.
6. German name of Transylvania, a historical region that is nowadays located in central Romania. Tisza, Maros, and the Danube are major rivers crossing the Carpathian Basin. The Author.
7. Latin meaning, without it. The Author.
8. Acronyms in this sentence are referring to various classes and grades of a wool grading system developed by Rudolph André.
9. French phrase, family resemblances. The Author.
10. Latin, by the fact alone.
11. Latin, forevermore or "to infinity". The Author.

Abbreviations

Hesperus	*Belehrung und Unterhaltung für Bewohner des Österreichers States*, Prag [Education and entertainment for residents of the Austrian State, Prague]
MAS	*Mährisch-schlesischen Gesellschaft zur Beförderung des Ackerbaues, der Natur- und Landeskunde* [Royal and Imperial Moravia-Silesian Association for the Furtherance of Agriculture, Natural Science and Knowledge of the Country] or *Ackerbaugesellschaft* [Moravian Agricultural Society]
Mittheilungen	Mittheilungen der kaiserreich-königlich Mährisch-Schlesischen Gesellschaft zur Beförderung des Ackerbaues, der Natur- und Landeskunde in Brünn. [Announcements of the Imperial-Royal Moravian-Silesian Society for the Promotion of Agriculture, Natural and Regional Studies in Brno]
ONV	*Oekonomische Neuigkeiten und Verhandlungen* [Economic News and Announcments]. Printed in Prague.
PTB	*Patriotisches Tageblatt oder öffentliches Korrespondenz- und Anzeiger-Blatt für sämtliche Bewohner aller kaiserreich-königlich Erbländer über wichtige, interessierende, lehrreiche oder vergnügende Gegenstände zur Beförderung des Patriotismus, Brünn* [Patriotic daily news or public correspondence and gazette for all residents of the imperial-royal hereditary lands about important, interesting, instructive or enjoyable objects for the promotion of patriotism, Brno]
SBS	*Verein der Freunde, Kenner und Beförderer der Schaftzucht, zur noch höheren, gründlichen Emporhebung dieses Oekonomie-Zweiges und der*

*darauf gegründeten, wichtigen Wollindustrie in
Fabrikation und Händel* [Association of Friends,
Experts and Supporters of Sheep Breeding
for the achievement of a more rapid and more
thoroughgoing advancement of this branch
of the economy and the manufacturing and
commercial aspects of the wool industry that is
based upon it] or *Schafzüchtervereinigung* [Sheep
Breeders' Society]

References

Almási, G. (2016) Faking the national spirit: Spurious historical documents in the service of the Hungarian National Movement in the early nineteenth century. *Hungarian Historical Review* 5:225–249.

Alston, P. L. (1969) *Education and the State in Tsarist Russia*. Stanford, CA: Stanford University Press.

Anderson, J. (1796) Letters and papers on agriculture, planting, etc. *Selected from the correspondence of the Bath and West England Society for the encouragement of agriculture, manufactures, arts and commerce* (4th edition), Vol. 8. Bath: R. Cruttwell.

André, C. C. (1795) *Der Zoologe, oder Compendiöse Bibliothek des wissenswürdigen aus der Thiergeschichte und allgemeinen Naturkunde*. Halle: Eisenach.

André, C. C. (1802) *Herr Hofrath von Geissler zu Hoschtitz, einier unser ersten Landwirte in Mähren*. Patriotisches Tageblatt 919–921.

André, C. C. (1804) *Anleitung zum Studium der Mineralogie für Anfänger*. Vienna.

André, C. C. (1815) Rede bey der ersten Eröffnung der vereinigten Gesellschaft des Ackerbaues, der Natur- und Landeskunde. Pages 93–111 in C. C. André (ed.), *Erster Schematismus der k. k. Mährisch-Schlesischen Gesellschaft zur Beförderung des Ackerbaues, der Natur- und Landeskunde*. Brno.

André, C. C. (1817) Abschrift des Zeugen-Verhörs in Betreff der grausamen That, welcher Elisabth v. Báthori, Gemahlinn des Grafen Franz Nádasdy beschuldiget wird. *Hesperus* 1611:241–256.

André, C. C. (1818) Terminologie für Woll-industrie. *Oekonomische Neuigkeiten und Verhandlungen* 302–343.

André, C. C. (1819) Erklärung des Herrn Grafen Emmerich von Festetics. Anmerkungen des Herausgebers. *Oekonomische Neuigkeiten und Verhandlungen* 4:26–27.

André, C. C. (1823) Preis von hundert dukaten. *Hesperus* 36:141–144.

André, C. C. (1826) Rudolph André starb zu Lischuwiz in Mähren den 9 Januar 1825. 32 Jahre alt. *Hesperus* 42:165–168.

André, R. (1812) *Oekonomische Neuigkeiten und Verhandlungen*. Pages 401–402. Review of the book by A. Thaer. Berlin: Handbuch der feinwolligen Schafzucht.

André, R. (1813) Review of the German translation of the French book by Tessier from 1811. Pages 281–283, 289–295, 297–300. *Oekonomische Neuigkeiten und Verhandlungen*. Berlin: Ueber die Schafzucht, insbesondere der Merinos.

André, R. (1815) *Anleitung zur Veredelung des Schafviehs. Nach Grundsätzen, die sich auf Natur und Erfahrung stützen.* Prague: JG. Calvé.

André, R. (1816) Anleitung zur Veredelung des Schafviehes. *Nach Grundsätzen, die sich auf Natur und Erfahrung stützen.* Prague.

André, R. (1818) *Kurzgefaßter Unterricht über die Wartung des Schafvieh's für Schafmeister und ihre Knechter faßlich eingerichtet.* Brno.

André, R. (1819) Meine Ansichten und Bemerkungen über organische Schwäche, besonders bei feinwolligen Schafen: veranlasst durch den Aufsatz des Herrn Grafen Emmerich von Festetics in Jännerheft 1819. *Oekonomische Neuigkeiten und Verhandlungen* 21(Beilage):161–162.

André, R. (1820) *Ideen über die Verwaltung landtäflicher Güter in Böhmen, Mähren und Oestreich: Ein Beitrag zur Darstellung der gegenseitigen Verhältnisse zwischen Gutsbesitzern, ihren Beamten und Unterthanen, so wie zur richtigen Würdigung des Wirthschafts-Beamten-Standes und des Besitzthums landtäflicher Güter.* Prague: F. Tempsky, JG Calvé.

Anon. (1774) Disciplina domestica (§24, Seite 24). *Vertheidigte Medietät und Landsässigkeit der Abtey Maximin bei Trier und ihrer im Erzstift gelegenen Güter: besonders der davon zu Lehen gehenden Mediatherrschaften Taben und Freudenberg.* Bayerische Staatsbibliothek: Abtei Sankt Maxim.

Anon. [D. Le.] (1800) Schafzucht. *Patriotisches Tageblatt* 47.

Anon. [probably C. C. André] (1809) Hesperus. *Ueber die Veredlung der Hausthiere* 94–101.

Anon. (1818) Wirksamkeit der Ackerbaugesellschaft in Brno. Schreiben an den Herrn Rath André in Brno. *Oekonomische Neuigkeiten und Verhandlungen* 297–304, 305–310.

Anon. (1818a) Ein neuer Schafzüchter-Congress in Ungarn. *Oekonomische Neuigkeiten und Verhandlungen* 329–330.

Anon. (1821) Ueber Kalender. *Literarisches Conversations-Blatt* 90: 357–360.

Anon. (1824) Constanz originaler Vollkommenheit in der Fortpflanzung. *Oekonomische Neuigkeiten und Verhandlungen* 29:231–232.

Bartenstein, B. (1837) Aeußerungen von Bartenstein über das von dem Herrn Professor Nestler bei dem Brünner Schaf-Züchter-Verein in Jahre 1836 aufgestellte und Thema: der Vererbungsfähigkeit edler Stammthiere. *Mittheilungen* 2:9–10.

Bartenstein, E., Teindl, F., Hirsch, J. and Lauer, C. (1837) Protokol über die Verhandlungen bei der Schafzüchter-Versammlung in Brno in 1837. *Mittheilungen* 201–205, 225–231, 233–238.

Bartosságh, J. (1837) A' Merinos, vagy selyembirka-fajok' elnevezéséről. *Társalkodó* 77:306–308.

Bateson, W. (1902a) Experiments with poultry. *Reports of the Evolution Committee of the Royal Society* 1:87–124.

Bateson, W. (1902b) Mendel's Principles of Heredity: A Defence. Cambridge, UK: Cambridge University Press.

Bateson, W. and Punnett, R. C. (1905–1908) Experimental studies in the physiology of heredity. Pages 42–60 in J. A. Peters (ed.), *Classic papers in genetics*. Englewood Cliffs, NJ: Prentice-Hall, 1959.

Baumann, C. (1785) *Nothwendige Anstalten zur Vermehrung, Verbesserung und Verschönerung der Pferd- Rindvieh- Schaf- Geiss- und anderer Thierzuchten ohne Ausartung*. Frankfurt-Liepzig.

Baumann, C. (1803) *Der Kern und das Wesentlische entdeckter Geheimnisse der Land- und Hauswirtschaft, zur bequemern Uebersicht und zum ausgebreitetern Gebrauch, mit der neunsten bewährten Versuchen und Nahrungsquellen, Liebhabern zum Handbuch gewidmet*. Brno: F.K. Siedler.

Bentley, S. M. (2009) *Friedrich Schiller's play: A theory of human nature in the context of the eighteenth-century study of life*. PhD thesis, University of Louisville, USA.

Bidinost, F., Roldan, D. L., Dodero, A. M., Cano, E. M., Taddeo, H. R., Mueller, J. P. and Poli, M. A. (2008) Wool quantitative trait loci in Merino sheep. *Small Ruminant Research* 74:113–118.

Blum, J. (1978) *The End of the Old Order in Rural Europe*. Princeton, NJ: Princeton University Press.

Blumenbach, J. F. (1781) *Über den Bildungstrieb und das Zeugungsgeschäft*. Göttingen.

Brem, G. (2015) Graf Imre Festetics—ein früher Pionier der Tierzucht und Genetik. Pages 69–75 in J. Seregi (ed.), *Kétszázötven év Ságtól Simaságig—1st Festetics Memorial Lectures 2nd Dec. 2014. S. 1.*

Brown, R. D. (1987) *Lucretius on Love and Sex*. Edited by E. J. Brill. Leiden.

Buffon, G-L. L. de. (1753) Le cheval. In: *Histoire Naturelle, générale et particulière, avec la description du Cabinet du Roi vol. IV. (Quadrupèdes I)*. Paris: De L'Imprimerie Royale.

Buffon, G-L. L. de. (1760) *Histoire Naturelle, générale et particulière avec la description du Cabinet du Roi vol. VIII. (Quadrupèdes V)*. Paris: De L'Imprimerie Royale.

Cartron, L. (2007) Degeneration and "Alienism" in early nineteenth-century France. Pages 1500–1870 in S Müller-Wille, H-J Rheinberger (eds.), *Heredity produced: At the crossroads of biology, politics and culture*. Cambridge, MA, USA: MIT Press.

Caspari, E. W. and Marshak, R. E. (1965) The rise and fall of Lysenko. *Science* 149:275–278.

Castoriadis, C. (1991) *Philosophy, Politics, Autonomy: Essays in Political Philosophy*. Edited by D. A. Curtis. New York, NY and Oxford: Oxford University Press.

Correns, C. (1900) G. Mendel's Regel über das Verhalten der Nachkommenschaft der Rassenbastarde. *Berichte der Deutschen Botanischen Gesellschaft* 18:158–168.

Cséby G. (2013) *Gróf Festetics György helye a magyar művelődéstörténetben, különös tekintettel a Magyar Minerva könyvsorozatra és a Helikon Ünnepségekre [Count György Festetics in Hungary's cultural history, with special regard to the book series "Hungarian Minerva" and the Helikon Festivities]*. PhD dissertation. Szeged: Szegedi Tudományegyetem.

Csíki L. (1971) *History of Our Institute from the Georgikon to Our Days*. Keszthely: Agricultural University.

Darwin, C. R. (1859) *On the Origin of Species by Means of Natural Selection, or the Preservation of Favoured Races in the Struggle for Life*. London: John Murray. p. 469.

Darwin, C. R. (1868) *The Variation of Animals and Plants under Domestication*. London: John Murray.

d'Elvert, C. (1870) *Geschichte der k.k. mäher.-chles. Gesellschaft zur Beförderung des Ackerbaues, der Natur- und Landeskunde, mit Rücksicht auf die bezüglihen Cultur-Berhältnisse Mährens und Österr*. Brno: Schlesiens.

de Vries, H. (1900) Sur la loi de disjunction des hybrids. *Comptes rendus de l'Académie des Sciences* 130:845–847.

Deák, Á. (2013) *Államrendőrség Magyarországon, 1849–1867. Akadémiai doktori értekezés [The State Police in Hungary, 1849–1867: A doctoral dissertation for the Hungarian Academy of Sciences]*. Budapest: MTA.

Deák, E. (2001) A tudományos élet és magyarországi kapcsolatai Cseh-Morvaországban a felvilágosodás korában [Scientific life and Hungarian connections in Bohemia and Moravia in the age of Enlightenment]. *Aetas* 16(3–4):29–45.

Diebl F (1829) Bemerkungen über die von Freiherrn v. *Witten hinsichtlich der verschiedenen Weizenarten geäusserten Ansichten. Mittheilungen* 177–179.

Diebl, F. (1835–1841) *Abhandlungen aus der Landwirtschaftskunde für Landwirthe, besonders aber für diejenigen, welche sich der Erlernung dieser Wissenschaft widmen*, Vol. 1–4. Brno.

Diebl, F. (1839) Ueber den nothwendnigen Kampf des Landwirthes mit der Natur, um die Veredlung und Erhaltung seiner Kultur-Gebilde. *Mittheilungen* 270–371.

Diebl, F. (1844) *Lehre von der Baum-Zucht überhaupt, und von der Obstbaumzucht, dem Weinbaue und der wilden – oder Waldbaumzucht insbesondere*. Brno.

Dixson, A. F. and Dixson, B. J. (2011) Venus figurines of the European Paleolithic: Symbols of fertility or attractiveness? *Journal of Anthropology*. Article ID 569120, DOI:10.1155/2011/569120.

Dobzhansky, T. (1937) *Genetics and the Origin of Species*. New York, NY: Columbia University Press. (2nd ed., 1941; 3rd ed., 1951)

Doyle, W. (1978) *The Old European Order, 1660–1800*. London: Oxford University Press.

Dunn, L. C. (1965) *A Short History of Genetics: The Development of Some of the Main Lines of Thought, 1864–1939*. New York, NY: McGraw-Hill. p. 65.

Ehrenfels, J. M. (1817) Ueber die höhere Schafzucht in Bezug auf die bekannte Ehrenfelsiche Race. *Belegt mit Wollmustern, welche die dem Herausgeber in Brno zu sehen sind. Oekonomische Neuigkeiten und Verhandlungen* 81–85, 89–94.

Ehrenfels, J. M. (1831) Fortsetzung der Gedanken des Herrn Moritz Beyer über das Merinoschaf. *Mittheilungen* 137–142.

Ehrenfels, J. M. (1837) Schriftlicher Nachtrag zu den Verhandlungen der Schafzüchter-Versammlung in Brünn, am 10. *Mai 1836. Mittheilungen* 1:2–4.

Elliott, P. and Daniels, S. (2006) The 'school of true, useful and universal science'? Freemasonry, natural philosophy and scientific culture in eighteenth-century England. *British Journal for the History of Science* 39:207–229.

Esparza, J., Schrick, L., Damaso, C. R. and Nitsche, A. (2017) Equination (inoculation of horsepox): An early alternative to vaccination (inoculation of cowpox) and the potential role of horsepox in the origin of the smallpox vaccine. *Vaccine* 35:7222–7230.

Fári, M. G. (2017) A liszenkóizmus előzményei, tündöklése, bukása és utóélete napjainkig a magyar növénygenetikában [Antecedents to Lysenkoism, its rise, fall and aftermath to date in Hungarian plant genetics]. *Debreceni Szemle* 25(2):148–169.

Festetics, E. (1815a) Híradás a juhtenyésztés jobbítását és pallérozását óhajtó hazafiakhoz [A call on patriots eager to improve and cultivate sheep breeding]. *Nemzeti Gazda* 10:145–147.

Festetics, E. (1815b) Aus einem Schreiben des Herrn Grafen Emmerich Festetics zu Güns in Ungarn [From a Paper by Count Imre Festetics of Kőszeg]. *Oekonomische Neuigkeiten und Verhandlungen* 69: 547–548.

Festetics, E. (1819a) Erklärung des Herrn Grafen Emmerich von Festetics. (Vergleichen Nr. 38., 39. u. 55., 1818). *Oekonomische Neuigkeiten und Verhandlungen* 2 (Suppl.):9–12.

Festetics, E. (1819b) Erklärung des Herrn Grafen Emmerich von Festetics. *Oekonomische Neuigkeiten und Verhandlungen* 3:18–20.

Festetics, E. (1819c) Weitere Erklärung des Herrn Grafen Emmerich Festetics über Inzucht. *Oekonomische Neuigkeiten und Verhandlungen* 22:169–170.

Festetics, E. (1820a) Bericht des Herrn Emmerich Festetics als Repräsentanten des Schafzüchter-Vereins im Eisenburger Comitate [A Report by Count Imre Festetics on Behalf of the Sheep

Breeders' Association of Vas County]. *Oekonomische Neuigkeiten und Verhandlungen* 4(19):25–27.

Festetics, E. (1820b) Äuserung des Herrn Grafen Festetics [A Statement by Count Imre Festetics]. *Oekonomische Neuigkeiten und Verhandlungen* 15(20):115–119.

Festetics, E. (1820c) Bericht des Herrn Grafen Emmerich Festetics über die vom Schafzüchter-Verein angeregten Versuche der Heckeisfütterung nach Petrische Methode [A Report by Count Imre Festetics on Experimenting with Petri's Method of Chopped Fodder, Proposed by the Sheep Breeders' Society]. *Oekonomische Neuigkeiten und Verhandlungen* 25(20):193–195.

Festetics, E. (1822) Über einen Aufsatz des Hrn. I. R. in 3ten Hefte des Jahrganges 1821. *Oekonomische Neuigkeiten und Verhandlungen* 92:729–731.

Freudenberger, H. (2003) *Lost momentum: Austrian economic development 1750s–1830s (Studien zur Wirtschaftsgeschichte und Wirtschaftspolitik).* Vienna: Böhlau Verlag.

Fuß, F. (1795) *Anweisung zur Erlernung der Landwirtschaft in Königsreich Böhmen.* Prague.

Gable E. (1996) Women, ancestors, and alterity among the Manjaco of Guinea-Bissau. *Journal of Religion in Africa* 26:104–121.

Gaissinovitch, A. E. (1990) C. F. Wolff on variability and heredity. *History and Philosophy of the Life Sciences* 12:179–201.

Gallois Duquette, D. (1983) *Dynamique de l'art bidjogo (Guinée-Bissau), Contribution à une anthropologie de l'art des sociétés africaines,* Lisboa, Instituto de Investigação Científica Tropical, p. 261.

Glass, B. (1947) Maupertuis and the beginnings of genetics. *Quarterly Review of Biology* 22:196–210.

Goethe, J. W. (1790) *Metamorphose der Pflanzen.* Gotha, Carl Wilhelm Ettinger.

Goldstein, R. J. (1983) *Political Repression in 19th Century Europe.* Totowa, NJ: Barnes & Noble Books.

Gordin, M. D. (2012) How Lysenkoism became pseudoscience: Dobzhansky to Velikovsky. *Journal of the History of Biology* 45:443–486.

Hagemann, R. (2002) How did East German genetics avoid Lysenkoism? *Trends in Genetics* 18:320–324.

Hankiss, E. (2017) *Quantum mechanics and the meaning of life (iASK Working Papers).* Kőszeg: Institute of Advanced Studies. https://iask.hu/wp-content/uploads/2017/09/wp_hankiss_170530_en.pdf?x51955

Hanson, A. M. (1985) *Musical Life in Biedermeier Vienna.* Cambridge: Cambridge University Press.

Harney, J. W. (2004) *Worthy is the Lamb: Pastoral symbols of Salvation in Christian art and music*. Thesis in Master of Arts, Skidmore College, USA

Haubner, P. (1857) Az állattenyésztésben a szülők befolyása az ivadékra. *[The influence of parents on their offspring in animal breeding]* Gazdasági Lapok 9:461–462, 509–510.

Henry C (1993) The Bijago arebuko (Guinea-Bissau). Possession cult and power objects. *Systèmes de pensée en Afrique noire* 12:39–64.

Herder, J. G. (2002) *Philosophical Writings*. Translated and edited by Michael N. Forster. Cambridge, UK: Cambridge University Press.

Herder, J. G. (2006) *Selected Writings on Aesthetics*. Translated and edited by Gregory Moore. Princeton, USA: Princeton University Press.

Herrera, M., Roberts, D. C. and Gulbahce, N. (2010) Mapping the evolution of scientific fields. *PLoS One* 5:ei0355.

Hempel, G. C. L. (1820) Ueber die Entstehung und die wichtigkeit der Verschiedenen sorten der Gatreidearten. *Oekonomische Neuigeiten und Verhandlungen* 21:163–164.

Hlubek, F. X. (1841) Nekrolog. *Moravia* 93:371–373; 94:375–378.

Hossfeld, U. and Olsson, L. (2002) From the modern synthesis to Lysenkoism, and back? *Science* 297:55–56.

Ihde A. J. (1956) The pillars of modern chemistry. *Journal of Chemical Education* 33:107–110.

Irtep [Petri] (1812) Ansichten über die Schafzucht nach Erfahrung und gesunder Theorie. *Oekonomische Neuigkeiten und Verhandlungen* 1–5, 9–16, 21–23, 27–28, 45–48, 60–61, 81–85, 91–92, 106–107.

Joravsky, D. (1970) *The Lysenko Affair*. Cambridge, MA: Harvard University Press.

Julius, N. H. (1846) *England's Mustergegängniss in Pentonville, in seiner Buuart, Einrichtung und Verwaltung, abgebildet und beschrieben*. Berlin: T.C.F. Enslin.

K. in Mähren [probably M. Köller] (1811) Ist es nothwendig, zur Erhaltung einer edlen Schafherde stets fremde Original-Widder nachzuschaffen, und artet sie aus, wenn sich das verwandte Blut vermischet? *Oekonomische Neuigkeiten und Verhandlungen* 294–298.

Kaposi, Z. (2016) Tudomány és gazdaság: Festetics Imre gróf birtokai és gazdálkodása a 19. század első harmadában [Science and economy: The estates and estate management of Count Imre Festetics in the first third of the 19th century]. *Közép-Európai Közlemények* 9:67–83.

Khrushchev, N. S. (2007) *Memoirs of Nikita Khrushchev, Volume 3: Statesman (1953–1964)*. Translated by G. Shriver. Pennsylvania, PA: Pennsylvania State University Press.

Klein, J., Klein, N. (2013) Solitude of a humble genius—Gregor Johann Mendel. Vol. 1, *Formative Years*. Berlin: Springer.

Köcker, M. (1809) *Auszüge aus Briefen des Herrn Oekonom Köcker, auf den Fürst.* Salmischen Herrschaft Raitz in Herbst 1808 an den Herausgeber des letzten. *Hesperus* 3:277–303.

Kodek, G. K. (2011) *Brüder, reicht die Hand zum Bunde: Die Mitglieder der Wiener Freimaurer-Logen (1742–1848).* Wien: Löcker Verlag.

Kroupa, Jiří. (1998). The alchemy of happiness: The Enlightenment in the Moravian context. In: Teich Mikuláš (ed.), *Bohemia in history.* UK: Cambridge University Press.

Kroupa, J. (2006) *Alchymie štěstí, Muzejní a vlastivědná společnost.* Brno.

Kuhn, T. S. (1962) *The Structure of Scientific Revolutions.* Chicago, IL: University of Chicago.

Kühndel, J. (1938) *Začátky zednářství na Moravě,* Olomouc.

Kurucz, Gy. (1990) "Az új mezőgazdaság" irodalma az egykori Festetics-könyvtár anyagában [The farming literature of "new agriculture" in the former Festetics family library]. *Magyar Könyvszemle* 106(1–2):32–44.

Lauer, J. E. (1841) Züge aus dem Leben verstorbener Gesselschaftsglieder. *Dr. Johann Karl Nestler. Mittheilungen* 41:319–321.

Lehleiter, C. (2014) *Romanticism, Origins and the History of Heredity (New studies in the age of Goethe).* London: Bucknell Universtiy Press.

Lerner, J. (1992) Science and agricultural progress: Quantitative evidence from England, 1660–1780. *Agricultural History* 66:11–27.

Lidwell-Durnin, J. (2019) Inevitable decay: Debates over climate, food security, and plant heredity in nineteenth-century Britain. *Journal of the History of Biology* 52:271–292.

López-Beltrán, C. (1994) Forging heredity: From metaphor to cause, a reification story. *Studies in History and Philosophy of Science* 25: 211–235.

López-Beltrán, C. (2004) In the cradle of heredity; French physicians and L'Hérédité Naturelle in the early 19th century. *Journal of the History of Biology* 37:39–72.

Lukács G. (2009) A Festetics-birtokok gazdálkodási és vezetési reformja a XVIII. század végén [Farming and management reform at the Festetics estates in late 18th century]. PhD dissertation, Keszthely.

Malinowski, B. (1932) *The Sexual Life of Savages in North-Western Melanesia.* London: Routledge & Kegan Paul.

Matalová, A. and Sekerák, J. (2004) *Genetics behind the Iron Curtain: Its Repudiation and Reinstitualisation in Czechoslovakia.* Brno: Moravian Museum.

May, A. J. (1963) *The Age of Metternich, 1814–1848.* New York, NY: Holt Rinehart and Winston.

McClelland, J. (1980) *State, University and Society in Germany, 1700–1914.* Cambridge: Cambridge University Press.

McIntosh, C. (1992) The rose cross and the age of reason. *Eighteenth-century Rosicrucianism in Central Europe and its relationship to the Enlightenment*. Leiden, the Netherlands: Brill.

Mendel, J. G. (1863) Bemerkungen zu der graphisch-tabellarischen Uebersicht der meteorologischen Verhältnisse von Brno. *Verhandlungen des naturforschenden Vereines in Brno* 1:246–249.

Mendel, J. G. (1864) Meteorologische Beobachtungen aus Mähren und Schlesien für das Jahr 1863. *Verhandlungen des naturforschenden Vereines in Brno* 2:99–121.

Mendel, J. G. (1865) Meteorologische Beobachtungen aus Mähren und Schlesien für das Jahr 1864. *Verhandlungen des naturforschenden Vereines in Brno* 3:209–220.

Mendel, J. G. (1866a) Versuche über Pflanzen-Hybriden. *Verhandlungen des naturforschenden Vereines in Brno* 4:3–47.

Mendel, J. G. (1866b) Meteorologische Beobachtungen aus Mähren und Schlesien für das Jahr 1865. *Verhandlungen des naturforschenden Vereines in Brno* 4:318–330.

Mendel, J. G. (1867) Meteorologische Beobachtungen aus Mähren und Schlesien für das Jahr 1866. *Verhandlungen des naturforschenden Vereines in Brno* 5:160–172.

Mendel, J. G. (1869) Ueber einige aus küstlicher Befruchtung gewonnen Hieracium-Bastarde. *Verhandlungen des naturforschenden Vereines in Brno* 26–31.

Mendel, J. G. (1870) Meteorologische Beobachtungen aus Mähren und Schlesien für das Jahr 1869. *Verhandlungen des naturforschenden Vereines in Brno* 8:131–143.

Mendel, J. G. (1871) Die Windhose vom 13. *Verhandlungen des naturforschenden Vereines in Brno* 9:229–246.

Mendel, J. G. (1879) Regenfall und Gewitter zu Brno in Juni 1879. *Zeitschrift der österreichischen Gesellschaft für Meteorologie* 14:315–316.

Mendel, J. G. (1882) Gewitter in Brno und Blansko am 15. *August. Zeitschrift der österreichischen Gesellschaft für Meteorologie* 17:407–408.

Morgan, T. H. (1919) *The Physical Basis of Heredity*. Philadelphia, PA: J. B. Lippincott.

Müller-Wille S, Rheinberger H-J (eds.) (2007) *Heredity Produced: At the Crossroads of Biology, Politics and Culture, 1500–1870*. Cambridge, MA: MIT Press.

Münch, Ernst. (1834). *Julilus Schneller's Lebensumriss und vertraute Briefe an seine Gattin und seine Freunde*. Leipzig: J. Scheible.

Murray, M. A. (1934) Female fertility figures. *Journal of the Royal Anthropological Institute of Great Britain and Ireland* 64:93–100.

Nestler, J. K. (1827) Ueber Gährungslehre, durch Bispiele aus dem praktischen Leben belegt, als Anleitung zum Denken über dieselbe.

Mittheilungen 33:257–262, 34:268–271, 35:276–279, 36:285–288, 38:
301–302, 39:309–311, 40:315 and 318.

Nestler, J. K. (1829) Ueber den Einfluß der Zeugung auf die Eigenschaften
der Nachkommen. *Mittheilungen* 47:369–372, 48:377–380, 50:394–398,
51:401–404.

Nestler, J. K. (1831) Neues aus der alten Zeit. *Mittheilungen* 9:71–72.

Nestler, J. K. (1837) Ueber Vererbung in der Schafzucht. *Mittheilungen*
34:265–269, 35:273 and 279, 36:281–286, 37:289–293, 38:300–303,
40:318–320.

Nestler, J. K. (1839) Ueber Innzucht. *Mittheilungen* 16:121–128.

Novotný, G. (2002) Christian Carl André a jeho synové Rudolph a Emil
Karl. Cesty k monografii o jejich působení na Moravě [Christian
Carl André and his sons Rudolph and Emil Karl. Pathways to a
monograph on their work in Moravia]. Pages 617–631 in Pocta Janu
Janákovi. *Předsedovi Matice moravské, profesoru Masarykovy univerz-
ity věnují k sedmdesátinám jeho přátelé a žáci [A tribute to Jan Janák: The
chairman of the Moravian Matice, the professor of Masaryk University,
devotes his friends and pupils to his 70s].* Brno: Matice moravská.

Nyáry, A., Ehrenfels, J. M. (1830) *Statuten für einen Schafzüchter-Verein des
Königreichs Ungarn.* Pest, Füskúti Landerer Lajos.

Oden, R. A. (1979) The Contendings of Horus and Seth (Chester Beatty
Papyrus No. 1): A structural interpretation. *History of Religions*
18:352–369.

Olby, R. (1987) William Bateson's introduction of Mendelism to England.
A Reassessment British Journal for the History of Science 20:399–420.

Orel, V. (1965) *Gregor Mendel, zakladatel genetiky: populárně vědecký sborník.*
Brno: Blok.

Orel, V. (1973) The scientific milieu in Brno during the era of Mendel's
research. *Journal of Heredity* 64:314–318.

Orel, V. (1974) The prediction of the laws of hybridization in Brno already
in 1820. *Folia Mendeliana* 9:245–254.

Orel, V. (1975) The building of greenhouses in the monastery garden
of old Brno at the time of Mendel experiments. *Folia Mendeliana*
10:201–208.

Orel, V. (1977) Selection practice and theory of heredity in Moravia before
Mendel. *Folia Mendeliana* 12:179–221.

Orel, V. (1978) Heredity in the teaching programme of Professor J. K.
Nestler (1783–1841). *Acta universitatis Palackianae Olomucensis, fac.
rer. nat* 59:79–98.

Orel, V. (1983) Mendel's achievements in the context of the cultural
peculiarities of Moravia. Pages 23–46 in V. Orel and A. Matalová
(eds.), *Gregor Mendel and the foundation of genetics.* Brno: Moravian
Museum.

Orel, V. (1989) Genetic laws published in Brno in 1819. *Proceedings of the Greenwood Genetics Center (South Carolina)* 8:81–82.

Orel, V. (1997) The spectre of inbreeding in the early investigation of heredity. *History and Philosophy of Life Sciences* 19:315–330.

Orel, V. (2005) Contested memory: Debates over the nature of Mendel's paradigm. *Hereditas* 142:98–102.

Orel, V. and Fantini, B. (1983) The enthusiasm of the Brno Augustinians for science and their courage in defending it. Pages 105–110 in V. Orel and A. Matalová (eds.), *Gregor Mendel and the foundation of genetics*. Brno: Moravian Museum.

Orel, V. and Peaslee, M. H. (2015) Mendel's research legacy in the broader historical network. *Science and Education* 24:9–27.

Orel, V. and Verbik, A. (1984) Mendel's involvement in the plea for freedom on teaching in the revolutionary year of 1848. *Folia Mendeliana* 19:223–233.

Orel, V. and Wood, R. J. (1981) Early development in artificial selection as a background to Mendel's research. *History of Philosophy of Life Sciences* 3:145–170.

Orel, V. and Wood, R. J. (1998) Empirical genetic laws published in Brno before Mendel was born. *Journal of Heredity* 89:79–82.

Orel, V. and Wood, R. J. (2000) Scientific animal breeding in Moravia before and after the discovery of Mendel's theory. *Quarterly Review of Biology* 75(2):149–157.

Pajkossy, G. (2006) Egy besúgó Pesten a reformkor közepén (1839) [An informer in Pest in the middle of the Reform Period]. *Aetas* 4:5–20.

Paleček, P. (2016) Vítězslav Orel (1926–2015): Gregor Mendel's biographer and the rehabilitation of genetics in the Communist Bloc. *History and Philosophy of the Life Sciences* 38, article 4.

Parry, C. H. (1806) Part II History of the Merino-Ryeland breed of the Author. In: *Communications to the board of agriculture; on subjects relative to the husbandry, and internal improvement of the country*. London: Bulmer and Co.

Penders, B. (2017) Marching for the myth of science: A self-destructive celebration of scientific exceptionalism. *EMBO Reports* 18:1486–1489.

Petri, B. (1813) Meine Ideen und Grunsätze über die Zuzucht der Hausthiere. *Oekonomische Neuigkeiten und Verhandlungen* 193–196.

Petri, B. (1827) Gegenbemerkungen in Bezug auf die Recenssion der zweiten Auflage meines Werkes "Das Ganze der Schafzucht" (bei Carl Schaumburg und Comp. in Wein 1825) durch Herrn. Staatsrath Thaer in 16. Bande der Möglinschen Annalen der Landwirtschaft, seite 507–556, verglichen mit mehreren in diesem Annalen erschienenen Aufsäzen des Hern. *Recensenten. Mittheilungen* 1:1–4, 5:31–37, 8:57 and 62, 11:84–87, 14:106–109, 17:130–134.

Pinto-Correia, C. (1997) *The Ovary of Eve. Egg, Sperm and Preformation.* Chicago, IL and London: Chicago University Press.

Poczai, P., Bell, N. and Hyvönen, J. (2014) Imre Festetics and the Sheep Breeders' Society of Moravia: Mendel's forgotten 'research network'. *PLoS Biology* 12:eiooi772.

Polanyi, M. (1958) *Personal Knowledge: Towards a Post-Critical Philosophy.* London: Routledge & Kegan Paul.

Pomata, G. (2003) Comments on Session III: Heredity and medicine. In *Conference: A Cultural History of Heredity II: 18th and 19th Centuries, Prepoint* 247, 145–152. Berlin: Max-Planck-Institute for the History of Science.

Pražák, R., Deák, E. and Erdélyi, L. (2003) *Széchényi Ferenc és Csehország: Levelestár [Francis Széchényi and Bohemia: Scientific and cultural correspondence, 1780–1816].* Budapest: Országos Széchényi Könyvtár, Gondolat Kiadó. (In Czech, German, and Hungarian)

Prod'homme, J.-G. (1961) Beethoven to Simrock, 2 August 1794. In E. Anderson (ed.), *The letters of Beethoven.* London: Macmillan.

Proudhon, P.-J. (2011) The federative principle and the necessity of reconstituting the party of the revolution. In I. MacKay (ed.), *Property is theft! A Pierre-Joseph Proudhon anthology.* Edinburgh: AK Press.

Purvis, I. W. and Franklin, I. R. (2005) Major genes and QTL influencing wool production and quality: A review. *Genetics Selection and Evolution* 37(Suppl 1):S97.

Röckel, J. (1808) *Pedagogische Reise durch Deutschland veranlasst auf allerhöchsten Befehl der bayerischen Regierung in 1805.* Dillingen.

Roe, S. A. (1981) *Matter, Life and Generation: Eighteenth Century Embryology and the Haller–Wolff Debate.* Cambridge: Cambridge University Press.

Roger, J. (1997) *The Life Sciences in Eighteenth-Century French Thought.* Edited by K. R. Benson, translated by R. Ellrich. Stanford: Stanford University Press.

Roll-Hansen, N. (2005) The Lysenko effect: Undermining the autonomy of science. *Endeavour* 29(4):143–147.

Ropolyi, L. and Szegedi, P. (2000) *A tudományos gondolkodás története: előadások a természettudományok és a matematika történetéből az ókortól a XIX. századig [The history of scientific thinking: Lectures on the history of natural sciences and mathematics from antiquity to the 19th century].* Budapest: ELTE Eötvös Kiadó.

Sagan, C. (1996) Interview with Charlie Rose, 27 May, 1996. https://speakola.com/ideas/carl-sagan-science-last-interview-1996

Salm, H. F. (1820) Fortsetzung des Auszugs aus dem Vertrage des Herren Präses Grafen Salm, *Direktors der Ackerbaugesellschaft. Oekonomische Neuigkeiten und Verhandlungen, Beilage* 5(19):33–34.

Salm, H. F. and André, C. C. (1814) An die Freunde der vaterländischen Industrie und der inländischen Schafzucht insbesondere. *Oekonomische Neuigkeiten und Verhandlungen* 113–114.

Sandler, I. (1983) Pierre Louis Moreau de Maupertuis—a precursor of Mendel? *Journal of History of Biology* 16:101–136.

Sandler, I. and Sandler, L. (1985) A conceptual ambiguity that contributed to the neglect of Mendel's paper. *History and Philosophy of Life Sciences* 7:3–70.

Sauvigny, G. (1962) *Metternich and His Times*. London: Darton, Longman & Todd.

Sedlářová, J. (2016) *Hugo Franz Salm, průkopník průmyslové revoluce: železářský magnát – mecenáš – sběratel – lidumil: 1776–1836 [Hugo Franz Salm, pioneer of the Industrial Revolution: Iron mogul – patron – collector – philanthropist: 1776–1836].* Praha: NPÚ, ÚPS v Kroměříži.

Sekerák, J. (2010) Mendel's discovery in the context of Moravian and world science. Pages 199–204 in L.Galuška, J. Mitáček and L. Novotná (eds.), *Poklady Moravy/Treasures of Moravia: Story of a historical land*. Brno: Moravian Museum.

Shan, Y. (2016) Exemplarising the origin of genetics: A path to genetics (from Mendel to Bateson). Dissertation, University College, London.

Simunek, M., Hoßfeld, U. and Wissemann, V. (2011) 'Rediscovery' revised—the cooperation of Erich and Armin von Tschermak-Seysenegg in the context of the 'rediscovery' of Mendel's laws in 1899–1901. *Plant Biology* 13:835–841.

Sinclair, J. (1832) *The Code of Agriculture* (5th edition). London: Sherwood, Gilbert and Piper. p. 83.

Sober, E. (2008) *Empiricism*. Pages 129–138 in S. Psillos and M. Curd (eds.), *The Routledge companion to philosophy of science*. New York, NY: Routledge.

Stomps, Th. J. (1954) On the rediscovery of Mendel's work by Hugo de Vries. *Journal of Heredity* 45(6):293–294.

Stumpf, J. G. (1785) *Versuch einer pragmatischen Geschichte der Schäfereien in Spanien, und der spanischem in Sachsen, Anhalt-Dessau etc.* Leipzig, Germany: Joh. Gottfried Müllerische Buchhandlung.

Svarez, C. G. (1960) Darstellung einiger besonders wichtiger Materien. Über die Ehe. In: Hermann Conrad, Kleinheyer Gerc (eds.), *Vorträge über Recht und Staat*. Köln, Germany: Westdeutscher Verlag-Köln.

Syniawa, M. (2006) Biographical dictionary of Silesian naturalists. *Upper Silesian Nature Heritage Center*. Katowice, Poland: Upper Silesian Nature Heritage Center.

Szabó T., A. (2009) "Valók gráditsonkénti lépegetése" (1818) és a "Természet genetikai törvényei" (1819) ["Gradualism" by Ch. Bonnet (1818) and "Genetic Laws of Nature" by Imre Festetics (1819): The emergence of the terms genetics, selection and evolution in Hungary]. *Nőgyógyászati Onkológia* 14:73–96. http://nokfolyoirat. hu/files/509.pdf

Szabó T., A. (2016) Korszakos felismerések és tévhitek a genetikában, Festetics Imre és a „genetika" fogalmi fejlődése kapcsán [Epochal insights and misbeliefs in the study of heredity regarding Imre Festetics and the birth of genetics]. *Kaleidoscope* 7:175–188.

Szabó T., A. (2017) Ursprung des Begriffs Genetik und seine Verwendung vor und nach Mendel. *Nova Acta Leopoldina* 413:65–79.

Szabó T., A. and Poczai, P. (2019) The emergence of genetics from Festetics' sheep through Mendel's peas to Bateson's chickens. *Journal of Genetics* 98, article 63. DOI:10.1007/s12041-019-1108-z.

Szabó T., A. and Pozsik, L. (1989) A magyar genetika első tudományos emléke. I. Festetics Imre (1819) a beltenyésztésről (A természet genetikai törvényei) [The first scientific records of Hungarian genetics: Imre Festetics on inbreeding (The Genetic Laws of Nature, 1819)]. *Tudomány (Scientific American Hungarian edition)* 12:45–47.

Szabó T., A. and Pozsik, L. (1990) A magyar genetika születése: Festetics Imre elgondolásai a beltenyésztésről és a "természet genetikai törvényeiről"—1819-ben (Brno – Brno). Festetics Imre születésének 225. évfordulójára [The birth of Hungarian genetics: Imre Festetics's ideas on inbreeding and "the genetic laws of nature," presented in 1819 in Brno/Brno. Commemoration on the 225th anniversary of his birth]. *Természet Világa* 121–122:50–60.

Teindl, F. J., Hirsch, J. and Lauer, J. C. (1836) Protokol über die Verhandlungen bei der Schafzüchter-Versammlung in Brno am 9. und 10. Mai 1836. *Mittheilungen* 303–309, 311–317.

Thaer, A. (1804) *Einleitung zur Kenntnisse der englischen Landwirtschaft.* Vol 3. Hannover: Gebrüder Halm.

Thaer, A. (1811) *Handbuch für die feinwollige Schafzucht. Auf Befehl des Königl. Preuss. Ministeriums des Innerns herusgegeben.* Berlin: Ville und Krammer.

Tschermak-Seysenegg, E. v. (1900) Ueber künstliche Kreuzung bei *Pisum sativum. Berichte der Deutschen Botanischen Gesellschaft* 18, 232–239.

Tschermak-Seysenegg, E. v. (1958) *Leben und Wirken eines österreichischen Pflanzenzüchters. Beitrag zur Geschichte der Wiederentdeckung der Mendelschen Gesetze und ihre Anwendung für die Pflanzenzüchtung.* Berlin: Verlag Paul Parey.

Vecerek, O. (1965) Johann Gregor Mendel as a beekeeper. *Bee World* 46(3):86–96. Reprinted in volume 92 (2015), issue 3.

Wallace, A. R. (1858) On the tendency of varieties to depart indefinitely from the original type. In C. Darwin and A. R. Wallace, On the tendency of species to form varieties; and on the perpetuation of varieties and species by natural means of selection. *Journal of the Linnean Society (Zoology)* 3:45–62.

Wang, Z., Zhang, H., Xang, H., Wang, S., Rong, E., Pei, W., Li, H. and Wang, N. (2014) Genome-wide association study for wool production traits in a Chinese Merino sheep population. *PLoS One* 9:0107101.

Weiling, F. (1968) F. C. Napp und J. G. Mendel—Ein Beitrag zur Vorgeschite der Mendelschen Versuche. *Theoretical and Applied Genetics* 38:144–148.

Weingart, P. (1999) Scientific expertise and political accountability: Paradoxes of science in politics. *Science and Public Policy* 26(3): 151–161.

Westfall, R. S. (1971) *The Construction of Modern Science: Mechanisms and Mechanics.* UK: Cambridge University Press.

Wilson, C. (1995) *The Invisible World. Early Modern Philosophy and the Invention of the Microscope.* Princeton, USA: Princeton University Press.

Wilson, E. O. (1999) *Consilience: The Unity of Knowledge.* New York, NY: Vintage Books.

Wilson, P. K. (2007) Erasmus Darwin and the "noble" disease (gout): Conceptualizing heredity and disease in Enlightenment England. In: S. Müller-Wille, H-J. Rheinberger (eds.), *Heredity produced: At the crossroads of biology, politics and culture, 1500–1870.* Cambridge, MA: MIT Press.

Wolfe, A. J. (2018) *Freedom's Laboratory: The Cold War Struggle for the Soul of Science.* Baltimore, MD: Johns Hopkins University Press.

Wolff, C. F. (1759) *Theoria Generationis.* Halle.

Wood, R. J. and Orel, V. (2001) *Genetic Prehistory in Selective Breeding: A Prelude to Mendel.* Oxford: Oxford University Press.

Wood, R. J. and Orel, V. (2005) Scientific breeding in Central Europe during the Early Nineteenth Century: Background to Mendel's later work. *Journal of History of Biology* 38:239–272.

Wurzbach, C. (1869) Cyrill Franz Napp. In: *Kaiserlich-königliche Hof- und Stattsdruckerei.* Vol. 20. Vienna: Biographisches Lexicon des Kaiserthums Oesterreich.

Index

Note: Locators in italics represent figures.